STORM WARNING

STORM WARNING

*Water and Climate Security
in a Changing World*

ROBERT WILLIAM SANDFORD

RMB

RMB | Rocky Mountain Books Ltd.
rmbooks.com
@rmbooks
facebook.com/rmbooks

Cataloguing data available from Library and Archives Canada
ISBN 978-1-77160-145-0 (pbk.)
ISBN 978-1-77160-146-7 (epub)
ISBN 978-1-77160-147-4 (pdf)

Printed and bound in Canada

Distributed in Canada by Heritage Group Distribution and in the U.S. by Publishers Group West

For information on purchasing bulk quantities of this book, or to obtain media excerpts or invite the author to speak at an event, please visit rmbooks.com and select the "Contact Us" tab.

RMB | Rocky Mountain Books is dedicated to the environment and committed to reducing the destruction of old-growth forests. Our books are produced with respect for the future and consideration for the past.

We acknowledge the financial support of the Government of Canada through the Canada Book Fund and the Canada Council for the Arts, and of the province of British Columbia through the British Columbia Arts Council and the Book Publishing Tax Credit.

Nous reconnaissons l'aide financière du gouvernement du Canada par l'entremise du Fonds du livre du Canada et le Conseil des arts du Canada, et de la province de la Colombie-Britannique par le Conseil des arts de la Colombie-Britannique et le Crédit d'impôt pour l'édition de livres.

Disclaimer: The designations employed and presentations of material throughout this publication do not imply the expression of any opinion whatsoever on the part of the United Nations University (UNU) concerning legal status of any country, territory, city or area or of its authorities, or concerning the delimitation of its frontiers or boundaries. The views expressed in this publication are those of the respective authors and do not necessarily reflect the views of the UNU. Mention of the names of firms or commercial products does not imply endorsement by UNU.

For Anna Warwick-Sears and her highly committed colleagues, board members and partners, who together work to make the Okanagan Basin Water Board one of the most relevant organizations of its kind in North America.

CONTENTS

The Critical Role of Science in Assuring Water and Climate Security in Canada

There was a time when Canada had a world-class reputation for building its prosperity on superbly clean and abundant water and outstanding water science and management. Recently, however, we have abandoned the fundamental principles that made us world leaders in water science and management and are now confronted with a perfect storm of water issues we could have prevented had we stayed on course. We face increasingly damaging industrial, mining and agricultural water abuse; troubling changes in local hydrology brought about by wetland drainage and inappropriate development in areas where we should be protecting source waters; and ever more damaging extremes of flood and drought brought about by climate changes to which we have contributed by changing the composition of the Earth's atmosphere. Despite clear evidence of the long-term damage we are doing to our own economy by ignoring the fundamental principles of sound water management, the public continues to tolerate dramatically reduced federal water science, monitoring and management capability. The circumstances that resulted in the premature ending of the Western Canadian Cryospheric Network (wc²n), the Improved Processes and Parameterization for Predictions in Cold Regions Network (ip3) and the Drought Research Initiative (dri) as the Canadian Foundation for Climate and Atmospheric Sciences concluded its funding of science in 2011 were very unfortunate for Canada and we are now trying to maintain, restore and build upon our scientific capacity through the nserc-funded Changing Cold Regions Network and other initiatives.

In the absence of strong federal commitment to water and water-related climate issues, problems that once received substantive national attention have been devolved to the provinces, which then download them onto local communities that have limited capacity to deal with them effectively. At this level, underfunded and fractured approaches to watershed management and disaster mitigation are made worse by uneven and inadequate flood and drought prediction and by transboundary water disagreements that are beyond the jurisdiction of municipalities to address. This perfect storm of abdicated responsibility has led to serious contamination of prominent transboundary water bodies. Lake Winnipeg, for example, now has algal blooms covering as much as 15,000 square kilometres, making it one of the most threatened major freshwater bodies in the world. In addition to the contamination threat, flood and drought crises along with related storm damage and forest fire impacts are now hammering every region of the country. These issues are restricting economic growth in Canada, limiting agricultural and energy production, destroying infrastructure and creating appalling water quality and human health conditions for First Nations and many other downstream communities. Declining federal leadership due to the loss of the Inland Waters Directorate, National Hydrology Research Institute, Freshwater Institute, Prairie Farm Rehabilitation Administration, Navigable Waters Protection Act and environmental impact assessment requirements, together with inadequate investment in weather, water quality and hydrological forecasting and observations has left Canada highly vulnerable to massive water crises that can cripple the national economy and that are already impoverishing regional economies.

We cannot afford the status quo; the economic losses due to the Prairie drought of 2000–2004 exceeded $4-billion and the Alberta–Saskatchewan–Manitoba floods of 2011–2014 exceeded $11-billion. The cost of one day of heavy rainfall in June 2013 in Toronto was just under a billion dollars. The Prairies and BC have been hit by drought yet again in 2015 and this time it restricted both food and oil production and even impacted the sport fishery. Poor international perception of water quality concerns is putting the ability to transport or export oil at risk. Contamination of Lake Winnipeg and Lake Erie from agricultural fertilizers has given them prominence as North America's main freshwater ecological crises. Undrinkable water puts First Nations on an infamous "third world" footing – unacceptable in a G7 country. Problems like this are accelerating rapidly.

The experience of Europe and the United States suggests that substantial economic, environmental and quality-of-life benefits can accrue from increasing national-level coordination and leadership in basin-scale water management based on integrated, intelligent data collection, strong science and coordinated severe weather and water prediction systems that provide reliable, ongoing decision support to governments at all levels. The United States has invested massively in a National Water Center to do just this. A co-operative Canadian approach could plan for, mitigate and reduce the impact of water crises due to floods, droughts and water contamination, but it would need the return of national interest in flood and drought science and forecasting, flood plain disaster reduction, climate change mitigation, water quality management, transboundary water management and science-based advice to agricultural producers, municipalities, watershed councils, industry, provinces and First Nations. Clearly we face tremendous water challenges in the future. Canadian hydrological and climate science stands ready to contribute to this, and Bob Sandford's book shows many ways in which we have delivered and still could deliver. It is with great optimism that we look forward to a future of co-operative approaches to water science and management in an era of big data, new observational technologies, powerful computer modelling capability and citizen interest in the application of science to a knowledge-based society of enlightened decision-making.

— John Pomeroy,
Canada Research Chair in
Water Resources and Climate Change,
University of Saskatchewan,
Saskatoon

CHAPTER ONE

COME HELL AND HIGH WATER

Scientific Truth and Consequences in Hollywood Disaster Movies

Anyone who has ever been tasked with inspiring the public imagination with an accurate interpretation of the relationship between water and climate will have quickly realized the great difficulty inherent in such a challenge. It is not that the public is not intelligent or interested. Many of them are both. The problem is that popular culture has gotten there first. As every scientist who has stood before the public will appreciate, we as a society – education notwithstanding – derive a great deal of what we know, and think we know, not from science directly but from print media, movies, television and the Internet. Unfortunately, that information is not always accurate.

Much of what people presume to know about dinosaurs, for example, comes from watching *Jurassic Park*. One of the reasons we so love such films is that they demonstrate what might be possible in the future if only we had enough imagination. Unfortunately, finding 200-million-year-old DNA intact in amber is not as easy as it looked in the movie. Nor is getting DNA out of an insect trapped in amber easy to do without contaminating it. Filling in large gaps in genetic strands with frog DNA are not likely to produce a viable genome, because the strands' constituents have not evolved together and in effect don't know how to talk to each other. Even if you were successful in implanting DNA into an appropriate egg medium, it would be difficult to overcome developmental problems associated with gene splicing. If finally you were able to hatch the eggs, there would be additional problems with compatibility with the radically different bacterial environments that exist today. There would also be challenges with humans replacing parents in behavioural instruction. Raising teenagers is one thing; raising velociraptors is quite another.

Because so many dinosaurs were predators, the number of them in Jurassic Park would also have required more room than they were provided in *Jurassic Park*. At 20 square miles, Isla Nublar was too small to provide adequate natural food supplies and prevent habitat destruction. The fact that the dinosaurs were hungry, however, did become evident in the film when they began to eat the central characters. Many readers will undoubtedly remember a highly symbolic, and for this viewer deeply satisfying, moment in the movie when a dinosaur ate the lawyer.

The world waited 22 years to see if these scientific problems would be solved in *Jurassic World*. But the genetic problems posed in the first movie were ignored. The chief geneticist in *Jurassic World* – the brilliant but ultimately evil Dr. Henry Wu – got around the problem of recreating exact genetic replicas of Jurassic dinosaurs by simply creating new creatures, the most prominent of course being the Indominus, a predator so ferocious that even the T-Rex needed the help of the park's resident Mosasaurus to bring it down. The problem of raising velociraptors did receive some attention, however, when Vic Hoskins, the head of security at the park, revealed he had a plan to weaponize them for military use. Once again the difference in climate during the Jurassic compared to the present was ignored. This remains a problem, at least scientifically.

As a consequence of undersea volcanism caused by sea-floor spreading associated with plate tectonics, the concentration of carbon dioxide in the atmosphere during the Jurassic was more than 2000 parts per million – five times what it is today. Predators aside, it was a much hotter world, in which humans would have found it very difficult if not impossible to survive.

CLIMATEGATE: A DENIALIST BESTSELLER

A great majority of people have also learned what they know, or think they know, about climate change not from science directly but from popular books, movies and made for television epics that interpret science in an entertaining and imaginative way. A great deal of current public interest in climate change emerges from what has been a highly public debate over whether or not it is actually happening, and if so, who or what is causing it. While these matters have been settled in the scientific community for some time, a high-profile and very profitable denial industry has come into existence to create a market for doubt. One of the vocal climate deniers portrayed in the opening scenes of a recently released James Balog film called *Chasing Ice* is a San Francisco television weatherman and talk-radio host named Brian Sussman.

In 2010 Sussman published *Climategate: A Veteran Meteorologist Exposes the Global Warming Scam*. The book is widely held to be one of the most vehemently anti-climate-change polemics ever written. While it can be painful for anyone versed in the hydro-climatic sciences to do so, it can be highly instructive and even useful to read such books and critically examine the arguments they put forward.

There is nothing subtle about this book. You know from the first sentence where the author is going. Sussman's acknowledgements do not mention a single scientist or climate expert. He nods to the usual suspects – his family, friends and literary agent – and then thanks the undisclosed sources who illustrated Al Gore's false science and the desperate climate scientists who tried to cover up the Climategate global warming scam for making his book possible.

This reviewer carefully examined the arguments in Sussman's book and after much consideration placed them into categories based on their principal themes. Of the 77 distinguishable arguments Sussman makes in *Climategate*, nine contest climate change for reasons of political ideology, three others on grounds of religion. Eleven more of his arguments are vicious *ad hominem* attacks on the character of climate scientists themselves, and 51 assail climate change on the grounds of incompletely reported, outdated, questionable or utterly erroneous science. To put it statistically, 74 of 77, or 96 per cent, of Sussman's arguments are objectively and scientifically unsupportable and therefore qualify only as personal opinion. This would not be a problem if Sussman didn't claim otherwise.

Despite protestations to the contrary, this book in no way proves that climate change is not a threat to the future of the United States and that we should continue, as he proposes, to use fossil fuels as long as we can if that is what it will take to perpetuate our prosperity.

If Mr. Sussman were to submit the arguments he has put forward in this book as a scientific hypothesis at any reputable academic institution, his thesis advisers wouldn't let him out of the room. Based on the quality of his arguments in *Climategate*, the best he might expect would be to receive a grade of 4 per cent, which qualifies on the grading scale as an F. In other words, he wouldn't graduate. Unfortunately, however, in our increasingly fragmented and argumentative society, the popularity of this book in some social and political circles will be the inverse of its scientific veracity. The more hateful it is in its condemnation of climate science and politicians that support action on climate change, the more popular it will become. That is what makes the book dangerous.

The main point this reviewer would like to make is that while *Climategate* may well represent a new standard for measuring the vehemence with which some interests oppose the idea of global warming, it would be a mistake to ignore this book. While it would be easy to dismiss the work by translating it into what this reviewer might call "the Sussman scale of denial," a new world standard against which anti-climate-science belligerence can be measured, that would not be an appropriate response. Rather, we should be thankful to Mr. Sussman for articulating his views and sharing them. His views tell science where it has to go to validate itself in the minds of those who think like he does. Somehow science has to build a bridge to those who hold views of this extreme hostility. If this is not done, then deeply held differences of opinion over climate change could literally tear our society apart.

However unpleasant it may be to have to refute arguments couched in the animosity Sussman expresses in *Climategate*, the ultimate credibility of the climate community resides in being willing to take the time to carefully, quietly and dispassionately separate fact from opinion. The first step in building such a bridge may be the clear identification of Sussman's arguments. As philosopher Christopher Lasch pointed out long ago, true dialogue demands that we enter imaginatively into our opponent's arguments, if only for the purpose of refuting them. Of course, there is always a chance that we ourselves will be persuaded by those we sought to persuade. But because argument is risky in this way, it is educational - hopefully for all of us.

Of the distinguishable arguments Sussman makes in *Climategate*, a number are political. The ideological unpinning of this book is loudly proclaimed in its opening sentence: "Global warming's story begins with a diabolical bastard named Karl Marx." Sussman's view is that climate change discourse has a direct link to covert communist intentions of undermining American democracy. He makes his views on this very clear in the book's introduction:

> The tentacles of Marxism have been steadily reaching into the United States for decades. Red-flag-waving propeller-heads in academia faithfully undermine American values in the classroom; useful idiots associated with a variety of non-governmental organizations work as tireless, agitating activists; and slimy liberal politicians craftily seek ways to undermine constitutional principles through, as stated by Lenin, political means. Working in lockstep with the unseen communist bureaucrats ensconced in the United Nations complex overlooking the East River in New York, this cabal of collectivists have discovered the ultimate tool to force social change upon first America, then the entire world: the very air we breathe. (p. xv)

Whipped into rage by what he considers an assault on his own extremist ideology, Sussman claims to hate all those he opposes. He reserves a special animus for academics and elites:

> An elite brigade of zealots has cleverly created a new political platform to carry out the collectivist goals of redistributing wealth and destroying personal liberty, utilizing something that Karl Marx himself never envisioned: the environment, or more specifically, the climate. And because the efforts are political, these

egalitarians are willing to utilize phony science as a terror tactic, in an attempt to force you to believe that your lifestyle is responsible for negatively altering the earth's atmosphere.

It's all a lie. (p. xvi)

Parsing the arguments

This book might have been far more useful to readers if its author had stuck to the facts and held off on the rhetoric, but that clearly is not his intention. If he had relied on peer-reviewed science and not merely his own opinions, this would have been a very thin pamphlet. The book might almost seem a caricature of ultra-right-wing thinking in America, so unrelentingly does Sussman hammer on ideological themes throughout. It is almost a case study of Poe's Law,* which posits that without a clear indicator of an author's intent, parodies of extreme views will, to some readers, be indistinguishable from sincere expressions of the views parodied.

Let's begin our summary of *Climategate*'s arguments with the ones that fall into the category of extreme political ideology:

- The first Earth Day in 1970 coincided with the 100th anniversary of the birth of Lenin. (p. 13)
- The entire global warming scam can be exposed through quotes attributed to either Marx or Lenin. (p. 63)
- The Environmental Protection Agency in the United States exists to curtail free enterprise and usurp private property. (p. 79)
- Climate scientists have equated global warming with the terrorist threat. (p. 105)
- The UN seeks to destroy America. (p. 143)

This assertion is supported by the following:

> Because the right to own property is a foundational principal [sic] of liberty and the pursuit

---

* According to Wikipedia, "Poe's Law" originated in 2005 from a forum comment posted by one Nathan Poe about creationism. It has become an Internet meme.

of happiness – and is antithetical to theories as-
cribed to Karl Marx – the overwhelming body
of the United Nations has always despised the
United States and has sought an effective means
of reshaping her. (p. 143)

- Sustainable development is just so much hooey; it is another
Marxist plot just like climate change. (p. 147)

This is supported by:

> [Sustainable development] provided the con-
> spiratorial template to assist in the creation of
> needless open-space reserves, forest preserves,
> marine reserves and supposed ecologically sen-
> sitive areas – all with the goal of removing mass
> acreage from future private development.

- The United Nations Framework Convention on Climate
Change was a secret communist plot to reduce the power,
prosperity and political influence of the United States. (p. 153)
- We desperately need an authentic conservative in the White
House. (p. 161)
- Nuclear power would be perceived as safe were it not for the
incompetence of the communists. (p. 171)

As mentioned earlier, three of Sussman's 77 arguments in
Climategate fall into the category of religious ideology. Naturally,
all three are highly charged:

- Climate change deniers are being treated like Copernicus and
Galileo by the ruling elite for their heretical views and for de-
fying a cult-like suppression of truth. (p. 126)
- Arm yourselves – cling to your gun and your religion. (p. 162)
- Humans should have dominion over all the rest of nature.
(p. 164)

A third category of Sussman's arguments would be the 11 personal
attacks on those he calls "climate propeller-heads." It is here that
his book suffers from the least pretense of objectivity. This reviewer

warns readers that the attack on climate science in this book is so relentless and personal that, if you know and respect some of his targets, it may be advisable to read Sussman's rant in small doses. As noted in the conclusion, Sussman may be using *ad hominem* attacks on the climate community to make a name and reputation for himself. It is difficult to understand the viciousness of the personal attacks in any other context. Some of these diatribes are multi-categorical and as such were hard to isolate into specific classes of argument. For example:

- DDT is safe and Rachel Carson is a fraud. (p. 6)
- Sussman adds to Argument 2 concerning the population bomb that never went off with a related point numbered Argument 3 on p. 152 in which he makes the point that environmentalists espousing population control are the agents of abortion and homosexuality. (p. 152)
- Michael Mann's hockey stick graph of temperature increases is an outright lie. (p. 35)
- James Hansen is clearly a Marxist and is lining his pockets to boot. (p. 47)
- One argument in this category could be construed largely as a bitter attack on "Little Al," a.k.a. Al Gore. Sussman's assertion here is that not one but two full chapters are required to properly characterize Mr. Gore's greedy, deceitful, hypocritical character and pro-communist, pro-abortionist, anti-Christian background and upbringing. (pp. 80–122)
- Any place that has Al Gore's books is a "lie-brary." (p. 100)

Only in a society that defends free speech would such an assault be permissible. Readers with legal backgrounds will observe that some of the sub-arguments border on slander. Other readers may simply laugh out loud. Many, however, have become weary of these kinds of tirades. The *Los Angeles Times*, for example, no longer publishes letters from climate change deniers, not out of censorship but out of a commitment to accurate reporting of factual information that is no longer in doubt. Readers are invited to judge Brian Sussman's comments for themselves.

- Environmental journalists and the propagandists they assist in the mainstream media are nothing but the feeding trough for sloppy science, brewed in cauldrons of bias and stirred by statist politicians and elitist social engineers. And sucking it up like chubby piglets are thousands of useful idiots. (p. 132)
- It is incomprehensible that a Republican presidential candidate of the reputation of John McCain would be sucked in by the climate hoax. (p. 135)
- Elites are evil and are trying to destroy America. (p. 186)
- Environmentalists are destroying the US, the last truly free country in the world. (p. 187)
- Smart energy grids are Marxist devices for controlling Americans. (p. 191)
- A CO_2 cap-and-trade program will destroy America. (p. 202)

The fifty-one of Sussman's points that are in the category of scientific arguments rely on outdated, partially accurate, inaccurate or completely fallacious assertions. Here is just a sample:

- Offshore oil wells are not an environmental hazard. (p. 10)

(Oops, the book came out before BP's "Deepwater Horizon" spill in the Gulf of Mexico in 2010.)

- When the Cuyahoga River in Cleveland caught fire in 1969, it only burned for 30 minutes and wasn't the big deal it was made out to be in terms of warning about water quality in American rivers. (p. 11)
- Climategate proves the fact that climate change scientists rely heavily on junk science. (p. 16)
- Climate scientists are becoming rich because of all the research funding they get. (p. 17)
- The predictions of a global cooling leading immediately to a new glacial period were false. (p. 20)

  (This one is true but Sussman does not explain how science later arrived at the same conclusion.)

- Higher atmospheric temperatures are not happening. They are being faked. (p. 44)

- We are not experiencing the hottest weather on record in the United States. Those temperatures occurred in the 1930s. (p. 56)
- Citing American novelist Michael Crichton as an expert, Sussman argues that there is not enough CO_2 in the atmosphere to cause global warming. (p. 67)
- Termites, cow rumination and elephant flatulence produce more CO_2 than humans. (p. 74)
- Fluctuations in solar radiation are not taken into account in models. (p. 76)
- S. Fred Singer is one of the world's greatest climate scientists. (p. 89)
- The threat of sea-level rise is utterly exaggerated. Coastal cities are not under threat. (p. 104)
- Sea levels globally are actually falling. Tuvalu is a tropical island mess being run by imbeciles who are using global warming as a shakedown operation the likes of which would make a Chicago community organizer proud. (p. 111)
- Arctic ice melt is not accelerating; percentage losses are attributable to arbitrarily increasing the area said to constitute the Arctic region. (p. 112)
- Arkansas Senator James Inhofe's list of 800 renowned scientists who oppose climate change clearly demonstrates there are no grounds for concern over catastrophic future warming. There are even American astronauts that don't believe it is happening. (p. 136)
- Coal is clean – soot, sulphur and nitrogen oxides are no longer a problem associated with coal in the United States. (p. 173)
- Air pollution is no longer a problem in the United States. (p. 180)
- There is no reason why we shouldn't expand fossil-fuel use. (p. 184)
- Sussman's book will help save America. (p. 208)

Summarizing the arguments

Sussman's conclusion offers five summary arguments meant to be the take-away from the book:

1. Humans are too small a force to ever affect anything as large as the global atmosphere.
2. Man-made global warming is the greatest scam in history.
3. We are living in an equivalent of 1938 in Germany and it would not be unreasonable to go to war over deeply held differences of opinion on climate change.
4. In the same way that those Germans who understood the threat Hitler posed to his own country should have spoken out in 1938, those who understand that climate change is a scam should speak out now against the climate fraud.
5. God will give the righteous the grace to do what is right.

Even if bona fide climate scientists may not be able to stand reading his book, Sussman does put some important points into relief. At a certain point of alienation, facts no longer matter, and accordingly Sussman ignores them. These include the following hard-won realizations:

1. The fact is that humans have become numerous enough to become capable of substantially altering the function of the Earth's biodiversity-based planetary life-support system. To understand this we need go no further than the realization that our impact on our planet's oceans.
2. The observed fact that our oceans really are acidifying.
3. The observed fact that temperatures are measurably rising and polar sea ice really is thinning and shrinking in extent.
4. The observed fact that climate warming feedbacks such as permafrost melt and methane release have now become clearly obvious.
5. The observed fact that predicted increases in climate variability are already evident worldwide.

To some, the content and style of *Climategate* will make its author out to be nothing more than a loudmouth crank and climate change bully. But this reader sensed there is something deeper wrong with this book. Despite his apparent personal outrage over matters climatic and his shoot-from-the-lip style, there is something studied about this tract that suggests careful positioning by expert public relations advisers. It is Sussman's calculated and carefully staged over-the-top rage that allows him to appeal so widely to like-thinking others. *Climategate* loudly and rudely proclaims what many others wish they could say were they not constrained by the courtesy, reason and respect for others that are the foundation of civility in contemporary society.

There are lessons in this for the climate science community. The first is that it is unwise to unnecessarily scorn or alienate knowledgeable people or even people who think they are knowledgeable on climate matters. They can become hateful enemies. To paraphrase William Congreve, *Climategate* demonstrates that hell hath no fury like a weatherman scorned. Brian Sussman has vigorously attacked contemporary climate science and what it has to say about how human activities have changed our planet's atmosphere and what that may mean for the future of our civilization. He has staked his ground and thrown down the gauntlet as if preparing for war. With this book Sussman has announced his intention to mount the climate debate equivalent of the invasion of the Sudetenland. And despite appearances, he is far from a lone soldier. His views are supported by many high-profile American and Canadian corporations, think tanks, media networks and moguls. How long this war will last with the rise of such forces as the Tea Party movement in the US and the deepening conservatism we have witnessed in Canada remains to be seen.

No one in science or politics should doubt for a moment that there are many more Brian Sussmans out there, many of them very well funded, and that each is trying to grow a following. Though the climate science community may wish to believe it is above engaging in such avoidably barbaric and unseemly activities, maybe scientists too should be preparing for war. Perhaps it really is 1938 all over again. Only this time we are not fighting against an outer tyranny poised to destroy our way of life. This time we may find ourselves facing a more terrifying enemy: a tyranny within that would destroy the world in an attempt to keep it the same.

All this said, it is only right and fair to acknowledge the fact that, despite Brian Sussman's aggressively contrarian disposition, there were arguments put forward in his book with which I largely agreed. All three are worth noting.

Though the first of these is not technically a climate-related argument, I certainly agreed with Sussman's opinion about the fans of the Minnesota Vikings. On page 30 he observes that by virtue of wearing horned hats and proclaiming that their quarterback is an ancestor of the Erik the Red, the beer-guzzling Vikings fans may well be certifiably crazy. Could be.

I also concurred with his argument on page 118 that the shrinking of Lake Chad is not caused by global warming, though he is only partly correct. The irrigation overdraft at work there is also a contributing factor to climate change impacts.

Another spot where Sussman and I are in full agreement concerns his argument on p. 105 that the movie *The Day After Tomorrow* does not accurately portray current climate science or future climate change scenarios. He is dead right. Climate disaster movies have in fact caused a lot of confusion about how climate actually works. So let's take a look at some of these films and see what they say about climate change. Let's also try along the way to show where the science has been bent out of shape for the purposes of entertainment, so that we can better understand what is really happening with respect to climate change.

DISASTER MOVIES: SCI-FI BECOMES CLI-FI

CLIMATE COOLING

The Day After Tomorrow was a smash hit in what has become a growing science fiction genre many call "disaster porn." Readers may prefer to follow author Naomi Klein on this and call the genre "cli-fi." *The Day After Tomorrow* was made in Toronto and Montreal in 2004 and, adjusted for inflation, is the highest-grossing Hollywood film ever made in Canada.

The film depicts catastrophic hydro-climatic effects in a series of extreme weather events that usher in global cooling which leads to a new ice age. After a massive drop in ocean temperatures in the North Atlantic, scientists conclude that the accelerated melt of polar ice has begun to disrupt the North Atlantic current, plunging the entire northern hemisphere into deep cold. The situation deteriorates when the resulting storm system develops into three hurricane-like super storms. In the eye of each of these storms air temperatures drop to as low 150° below zero and everything the air touches instantly freezes. This is scary stuff but is it likely to happen? Not anytime soon and certainly not at anything like the rate of cooling depicted in the film.

What would be required to cause the shutting down of the North Atlantic current would be a disruption in what is called the Atlantic Meridional Overturning Circulation, or AMOC. Sometimes referred to as the "global ocean conveyor belt," the AMOC is an important component of the global atmosphere–ice–ocean–climate system. This circulation system redistributes heat between polar and equatorial regions and among the Earth's ocean basins and atmosphere. It also contributes to the global ocean's sequestration of carbon generated through human activity. While there was genuine concern in the 1990s that the rapid melt of the world's great ice caps and ice sheets might pour enough cold freshwater into the Atlantic to disrupt the global ocean conveyor belt and bring about abrupt climate cooling in the northern hemisphere, later research suggested that the chances of this happening in this century were low. Further research results in 2014, however, revived concern over the apparently very real possibility of rapid ocean current disruption. All bets may be off if we continue to alter the composition of our atmosphere, in which case the science in *The Day After Tomorrow* might not appear nearly as laughably improbable to future generations.

SEA-LEVEL RISE

The prospect of rapid melt of the world's ice caps, ice sheets, icefields and glaciers is a theme that has been well explored by Hollywood. One of the truly frightening climate change scenarios is rapid and extreme sea level rise. The film that tried to scare us the most on this topic is Kevin Costner's 1995 epic *Waterworld*. The setting of the film is several centuries in the future. It appears that at the beginning of the 21st century, the polar ice caps began to rapidly melt and the sea level began to rise hundreds of metres, submerging every continent and turning Earth into a ball of water. Remaining human populations were subsequently scattered across the ocean in isolated communities consisting of artificial islands and barely seaworthy boats. The plot of the film centres, as the film's website explains, on an otherwise nameless antihero, "The Mariner," a drifter played by Kevin Costner who sails all over the Earth in his trimaran. Is this likely to happen? Not a chance. The first problem with the movie – at least in scientific terms – is that even if you melted all the ice on the planet, there would not be nearly enough water to submerge the continents.

Scientists have modelled what North America would look like if all the ice in the world melted. While you certainly might want to pay attention if you were living in California, the Caribbean, southern Texas, Louisiana, Florida, the Carolinas, New York, Boston or Halifax, everything would still be fine in the interior of the continent. Even in the unlikely event that the West Antarctic Ice Sheet melts, with mean sea level rise in the order of metres instead of centimetres, the consequences would be serious certainly but it would still not be the end of the world. A far more immediate climate-related threat is the potential effect of sea ice melt on ocean acidification, which is now occurring at a rate 1000 times faster than occurs naturally. Because of this, the world's oceans are now 30 per cent more acidic than they were around 1800. By 2050 our oceans are expected to be warmer and 150 per cent more acidic than they were two centuries ago, with cascading effects on all marine ecosystems. By mid-century we can expect sea levels to rise and land surfaces to be warmer. So let's see now what the movies have done with heat.

HEAT WAVES

Heat waves in themselves have not been the subject of any high-profile Hollywood disaster epics. But because they can be devastating and drive people to the limits of their physical and emotional endurance, heat waves have been the backdrop of a number of movies, including Spike Lee's Do the Right Thing. The plot of this 1989 film develops around racial tensions that explode into a riot in an urban American neighbourhood during a heat wave.

The plot of Do the Right Thing is hardly implausible. Heat waves annually kill people by the thousands in concentrated urban areas around the world. During four months in the summer of 2003, 70,000 people in 12 European cities died of heat-related ailments.

Temperature records were broken all across Europe again in 2006 and 2015. This problem is not going away. Rising levels of urban heat are now seen to constitute the single greatest climate-related threat to human health globally. Urban heat waves now account for more deaths per year globally than all other forms of extreme weather events. We don't have to wait for the future for this. It's happening now.

Because of the heat-island effect of dark-surface solar absorption, combined with waste heat from vehicles, industries, buildings and even people themselves, temperatures in cities can be as much as 12°C higher than in the surrounding countryside. With what would be only moderately warm temperatures in surrounding rural areas, we have discovered that big cities can create their own heat waves.

Heat-wave fatalities underscore the limitations of the human body to withstand extreme heat, which explains why we probably wouldn't have done very well in the Jurassic. It is true that we can, as a species, survive surprisingly high temperatures, in part because our metabolism operates at 37°C, which is usually considerably warmer than most outdoor temperatures. Above 37°, however, the body must cool itself through evaporation in the form of perspiration. Unfortunately for the very young, the very old and those with health problems, the human heart has to work harder in hot weather to distribute water for cooling through perspiration loss. Keeping up with this loss requires constant hydration. That is why most of the deaths during a heat wave happen at night. Sleeping people can't hydrate, so their core body temperature rises and their organs shut down or their heart stops.

One of the major reasons why big cities are so much warmer now than in the past is the loss of trees and other vegetation to development. Trees not only store heat-absorbing water, their canopies create shade that moderates high temperatures. Once the heat-island effect is generated it is not easy to reverse. As resilience to climate extremes becomes a design imperative, cities may have to make extensive changes, especially to their land- and waterscapes. The good news here is that we don't need international climate accord to make these changes. The power to do so already resides in the hands of those who live in vulnerable cities.

But urban heat waves are not the only threat. What about drought on farms?

DROUGHT

More recent science fiction films utilize deep and persistent drought and its impact on food supply and water security as a means of generating plot. In *Interstellar*, for example, the world's food supply is threatened by storms and disease to the extent it becomes necessary to explore the depths of space and time for an alternative, Earth-like home.

In *Mad Max: Fury Road* the drought and limited water supply that have left the planet barely habitable lead to a more violent, dystopian world than the one that emerges in *Interstellar*. It is a terrifying place, but they do drive interesting vehicles.

CATASTROPHIC STORMS

The hands-down all-time worst of the B movies about climate disaster has got to be the hilariously campy *Sharknado*. The premise of this biggie is that climate change is going to cause more powerful storms, which will have unexpected consequences. In *Sharknado*, the complete surprise is a massive storm off the coast of Mexico that forms a tornado at sea so powerful that it sucks 20,000 sharks up into the sky. And of course the storm hits Los Angeles with all its fury. The sharks fall out of the sky and in the ensuing flood take over the city, attacking people in their Beverly Hills homes, occupying their swimming pools and shooting out of manhole covers to eat unsuspecting pedestrians whole.

This flick was so bad they just had to make a sequel. As the second opus begins, the heroine of *Sharknado* the First, April Wexler, and the fearless Fin Shepard are on a flight approaching New York when the pilot announces unexpected turbulence. April looks out the window of the 747 and notices a shark stuck to the wing. Inexplicably sharks are soon inside the plane and one bites off one of April's hands. I should have explained that April is on the way to New York to promote her new book, *How to Survive a Sharknado*. I should also point out that weather forecasters were already predicting a monstrous storm. (In New York, evidently, "shark happens.")

April and Fin, of course, save the city after narrowly missing being crushed by the head of the Statue of Liberty which rolls down a street that soon becomes a river of man-eating sharks. Interesting plot – but the science is, well, questionable. If you are scientifically or rationally oriented, it would not be unreasonable to drink heavily before watching it.

Sharknado 3: Oh Hell No! escaped to television during the summer of 2015. "Three times the shark, three times the nado!" Aficionados might not think it possible, but despite higher production values it was a worse film than the other two. This reviewer loved it. In this writer's mind this epic redeemed itself with the single grain of scientific truth it offered when at the outset of a sharknado strike on his military base the commanding officer pointed out that the science of "bio-meteorology" was still in its infancy. Yup, that's true. Even Brian Sussman might agree with that.

A much more credible but still wildly exaggerated film on extreme weather events was a four-hour television mini-series produced in 2005 called *Category 7: The End of the World*. The premise of this riveting film was that storms of unprecedented ferocity are not just battering but destroying cities around the world. While television evangelists proclaim the end times, a newly charged and powerfully motivated FEMA, the US Federal Emergency Management Agency, is determined to find out what is triggering extreme weather disasters of a magnitude never witnessed before. With the help of renegade scientists, and of course the Air Force, it is discovered that urban heat islands are shooting thermal columns upward into the mesosphere, creating openings down which the cold of near space is brought to the surface of the Earth, causing never before witnessed Category 7 storms. But just as a Dick Cheney figure is sucked into the whirlwind and the White House is destroyed by the super-storm, the day is saved. The solution of course is to shut down all of the thermal power stations around Washington, DC.

Besides saving FEMA's reputation, *Category 7* contained one of the best lines of any of the classic cli-fi movies. Says the renegade scientist subhero to the beautiful and very intelligent scientist subheroine: "You are an egghead, but deep down you have a redneck heart." "Stop pussyfooting around and kiss me," she fires back. Evidently we can still expect hot stuff in a hotter world.

TORNADOS

Twister is a 1996 American disaster drama starring heartthrob Helen Hunt and Bill Paxton as storm chasers researching tornadoes. They are working to perfect a data-gathering instrument they have named "Dorothy" which they are trying to release into the funnel of a tornado during a twister outbreak across Oklahoma. They are of course competing with bad guys in the form of another, better-funded team with a similar device. The researchers finally get Dorothy launched but the climax of the film is when they find themselves chained to an irrigation pipe staring up into the vortex of a Category F5 tornado.

The plot is a dramatized view of research projects like VORTEX in the United States. The "Dorothy" device used in the movie is copied from real-life technology called the "TOtable Tornado Observatory," or TOTO, that was used in the 1980s by the National Severe Storms Laboratory, a National Oceanic and Atmospheric Administration research facility at the National Weather Center in Norman, Oklahoma.

Twister was gripping, but it was nothing compared with *Into the Storm*, which came out during the summer of 2014. Computer graphics have come a long way in 20 years, and so has our understanding of tornadoes. Readers are urged to see this one in a movie theatre with a big screen and big audio if they can. There are moments in this movie when you feel like you are being sucked from your seat right into the storm.

Interestingly, the tension in this film is not between teams of funded versus unfunded researchers but between scientific data and intuition in predicting where tornadoes will strike. All this is forgotten, however, when the heroes go into the storm. There is a lot of excellent scientific interpretation related to how tornadoes form and how they might act in a more energetic atmosphere, but as seems to happen with most Hollywood epics, the producers just couldn't resist going over the top at the end.

So where do we stand presently in terms of understanding tornadoes? In terms of absolute counts, the United States leads the list, with an average of over 1000 tornados recorded each year. Canada is a distant second, with around 100 per year. But trends are difficult to establish. Because tornadoes are short-lived events and we have such limited and unreliable data from sparsely populated regions where they typically occur, it is difficult to know for certain whether or by how much the intensity and frequency of tornadoes is increasing. One thing we do know, however, is that one of only nine Category F5 tornadoes to have occurred in North America between 1999 and 2011 touched down near Elie, Manitoba, on June 22, 2007. Though there was no loss of life, the F5 at Elie does remind us that Canada has its own Tornado Alley, extending across southern Ontario and picking up again in southern Manitoba and Saskatchewan before doing the twist around Edmonton, Alberta. Recently, however, tornado warnings have even extended into the mountains and as far north as the Yukon.

Some researchers think we are entering a new era of Earth-system disruption, called the Anthropocene. If that is the case, the greater apparent incidence of tornadoes has already provided an iconic image of what the era might look like. It is a picture both scary and hopeful. On Saturday, July 5, 2014, a young Saskatchewan couple chose to have their wedding photographs taken on a rural road outside of Regina. As photographer Colleen Niska began composing her shots a tornado formed in the distance behind the wedding couple and later blew down and flooded their reception tent. As the bride later wrote on Facebook: "We captured a once in a lifetime moment and feel so lucky to have it as a memory of how in such chaos you always have each other!"

If this Anthropocene is the brave new world we are all now headed for, it's good to know there will still be love and hope in it.

CLIMATE INSTABILITY: THE SCIENCE SO FAR

So, all the disaster movies and prairie wedding "photo bombs" aside, what is *really* happening out there? What can science tell us about all this that we are not getting from newspapers, television, the Internet or blockbuster Hollywood movies? You don't need to be a meteorologist to figure out that our weather is all over the place. Rain storms, ice storms and snow storms are paralyzing our transportation and electricity distribution systems. Both high and low temperature records are being broken everywhere. Cold snaps are persisting; snow is falling in places and in volumes seldom witnessed before; flooding is occurring widely. But even with all this obvious evidence right before our very eyes, Canadians continue tiptoeing around the climate change issue.

Science has begun to illuminate some of the linked hydro-climatic effects that researchers have identified as threats to the stability of our weather systems. One of the most significant recent advancements in hydro-meteorology is the realization of the global importance of the refrigerating influence of polar sea ice. Three factors related to the extent and duration of Arctic sea ice have emerged as having great significance to every one of us alive today. The first, as already noted, is the potential effect of sea ice melt on ocean acidification. Second is the amount of heat required to melt sea ice. The third critical effect is the profound influence of Arctic cold on the stability of weather conditions farther south.

POLAR ICE AND THE JET STREAM

Polar ice is now seen as a thermostat that governs major weather patterns globally. It is feared that the decrease in the extent and thickness of sea ice could be the parameter that is feeding all of the increases that are causing concern over climate disruption. There has been a 75 per cent decrease in the volume of Arctic sea ice in the past 35 years. The loss of sea ice allows solar heat to warm the Arctic Ocean, creating a feedback which melts more ice. This is why the Arctic is warming two to three times faster than the rest of the world, in all seasons.

The loss of Arctic sea ice and the diminution of the extent and duration of snowcover in the northern hemisphere are reducing the temperature difference between the colder, thinner polar atmosphere and the warmer, thicker atmosphere to the south. It is this gradient in temperature between the two regions that largely defines the behaviour of the jet stream. Observations of the jet stream have revealed that warmer Arctic sea surface and atmospheric temperatures do not automatically translate into warmer weather. The influence of warmer Arctic air causes the west winds created by the jet stream to become weaker. As researcher Jennifer Francis (see chapter ten) puts it, "Weaker westerly winds are wavier." It is this waviness of the jet stream that creates weather at mid-latitudes. The result is a slower, more sinuous jet stream with fingers that stretch farther north. The slower, wavier jet stream creates longer-lasting weather patterns. An example of this was the southward bend in the jet stream that brought record cold to much of eastern North America during the winter of 2013-14. In a uniformly warmer and therefore more turbulent atmosphere, both warm and cold fronts end up and persist in places in the mid-latitudes where they were not common in the past.

As the jet stream slows and the waviness becomes wider, weather patterns persist longer and do things we don't expect. When we look at extreme weather events such as heat waves and heavy rain and snow, we usually see very large jet stream deviations. One of the many things a slower, wavier jet stream can do is disrupt the polar vortex.

THE POLAR VORTEX

Much has been made lately in the media of the role of the polar vortex. There are actually two elliptical or circular polar vortices, one usually over the Arctic and one over the Antarctic in their respective winters. The latter vortex is very stable, but the Arctic one is more variable. It is influenced from below by the changing Arctic sea ice cover and from above by warming in the stratosphere. Recently both factors have weakened the vortex. A weak vortex is much more susceptible to disturbance at its periphery by the polar jet stream. The effects of such disturbances were made coldly obvious in North America during the winter of 2013-14. Instead of one consolidated vortex over the Arctic, four lobes developed, with the cold circulations carried south in the troughs of the jet stream. One of these lobes descended and lingered over central and eastern North America for most of the winter, making it one of the most miserable in memory in Manitoba and southern Ontario.

What we are also seeing is that changes in atmospheric circulation patterns are pushing major subtropical storm tracks toward the poles, often causing floods of magnitudes we are poorly equipped to manage.

What we might derive from all this is that it appears that the Arctic is more of a bellwether of climate change and disruption than we thought. While we may not have perceived it as such, what is happening hydrologically in much of the rest of Canada mirrors what is happening in the Arctic. Warming is causing the post-glacial hydrological wealth of Canada to change form. In the Arctic, sea ice is melting, permafrost is thawing and snowcover is diminishing in both extent and duration. Warming is also occurring in Canada's western mountains. Some 300 glaciers have disappeared over the last century in the Canadian Rockies alone.

The water is not disappearing, however. Water doesn't do that. The warming generated by changes in the composition of our atmosphere is being absorbed by water. This warming increases evaporation. More liquid water is moving to a different place in the hydrosphere, where it may not be available for our use when we want it – as in the case of extreme drought – and where under certain circumstances the warming atmosphere's capacity to carry more water can also cause a lot of damage through heavier than normal rains and the resultant flooding. What also appears to be happening is that the global hydrological cycle is accelerating. Warmer temperatures are causing changes in the rate and manner in which water moves through the cycle. The phenomenon is summed up in an algorithm called the Clausius–Clapeyron relation. It is one of the most basic principles in atmospheric physics, and what it posits is very simple: warmer air holds more water. The amount of water the atmosphere can hold increases by about 7 per cent per degree Celsius, or about 4 per cent per degree Fahrenheit.

ATMOSPHERIC RIVERS

We are also witnessing things we either haven't seen or weren't able to recognize as such before, such as the currents of water vapour aloft called atmospheric rivers. Atmospheric rivers have likely existed for an eternity but only now because of satellite remote sensing capacity do we know of their existence and dynamics. These corridors of intense winds and moist air can be 400–500 kilometres across and thousands of kilometres long. They can carry the equivalent of 10 times the average daily discharge of the St. Lawrence River. We have discovered recently that atmospheric rivers derive their energy from the temperature gradient between the poles and the tropics. Their intensity also derives from the Clausius–Clapeyron relation in that the warmer the air, the more water atmospheric rivers can carry.

Atmospheric rivers produce flooding of the magnitude we saw in Australia and Pakistan in 2010 and may have influenced flooding on the Canadian prairies in 2011. We know for sure, however, that an atmospheric river broke rainfall records by 87 per cent in the Kootenay–Columbia River region of British Columbia in 2012.

It was 2013, however, that really demonstrated, in Canada at least, that we may in fact have crossed over an invisible threshold into a new hydro-climatic regime in parts of the country. On the Canadian prairies, 2013 was interesting because despite a record late-season snowpack, there was no flooding in the spring. Most people have forgotten why.

UNPREDICTABILITY: THE NEW NORMAL

Why flooding didn't occur in 2013 was a subject of the annual Red River Basin Commission conference that was held in January of the following year in Fargo, North Dakota. At that conference I witnessed an extraordinary spectacle, one I would give anything to see happen in Canada. Imagine this: federal scientists with the National Weather Service had leave to give a public presentation on why they *failed* to make an accurate flood prediction in the spring of 2013. At the time, government scientists in Canada were not permitted to speak publicly at all unless accompanied by departmental media handlers. The US presentation demonstrated the public value of open scientific exchange.

Researchers reported that the snow-water equivalent in the April 2013 snowpack was so abnormally high that no one had experienced such conditions before. It seemed almost incomprehensible that record flooding could be avoided. But to the utter amazement of the National Weather Service, it didn't happen. Why? Because a series of other, equally unlikely extremes came into play that slowed the melt of the record late-season snowpack. Near-record cold weather slowed snowmelt, while timely thaw and warm deep-soil temperatures accommodated twice the normal rate of absorption.

The reason the National Weather Service offered this presentation was because they wanted to make sure everyone in the region clearly understood that the absence of flooding in 2013 did not in any way mean that hydro-climatic regimes in the Red River Basin had somehow returned to what had historically been perceived as normal. In this basin, the researchers noted, we face a growing number of "no analogue" situations where we are facing conditions we haven't seen before. The National Weather Service wanted the US public to know that we have entered a period in which more frequent extremes in hydrological conditions now make it impossible to use past experience as a guide to prediction.

While Manitoba may have dodged a bullet in 2013, not everyone else did. On the evening of Wednesday, June 19, rain began to fall where I live in Canmore, Alberta, initiating the largest single natural disaster in the history of this country. What was interesting about this flooding was that in terms of severity it was calculated by Dr. John Pomeroy of the University of Saskatchewan to be but a 1 in 45 year event associated with the spring melt of the mountain snowpack. In other words it was well within the natural variability experienced over the last century in Alberta. The communities that were hit by the flooding had clearly overbuilt in flood plains and were not adequately warned of the impending disaster. But the flooding in Canada that summer didn't stop with the disaster in southern Alberta. It continued widely and assumed a different character as it went. The further flooding that occurred that summer was associated not with spring melt but with highly charged heavy rainfall events. On July 8, just over two weeks after the flooding at Calgary, Toronto was hit by a rainstorm that caused over a billion dollars in insured losses alone. Disasters that followed that summer demonstrated, unfortunately, that the flooding in Alberta and Ontario in June and July of 2013 was nothing compared to what the atmosphere is capable of delivering as a result of changing hydro-climatic conditions.

What happened in Russia later that same summer is almost

beyond imagination. It verges on science fiction. The weakening of the European jet stream caused by reduced snow and sea ice cover led to the creation of a heat dome in northern Siberia. In July hundreds of wildfires broke out that were so hot they melted the permafrost beneath the burning forests, creating methane releases from the thawing tundra that added fuel to the fires. Then, in early August, in the midst of what was coming to resemble a virtual firestorm, three atmospheric rivers collided over the region and within four days created a flood that covered a million square kilometres.

Colorado too saw flooding in 2013, and the fact that it occurred in September, independently of spring melt, following severe drought and one of the worst wildfire seasons in history, is alarming in its own right. The storm demonstrated that subtropical storms are indeed tracking poleward and that the atmosphere is carrying more water. The state climatologist of Colorado, Nolan Doesken, noted that the storm "shattered all records for the most water vapor in the atmosphere."

Taking Colorado's experience together with Alberta's and Ontario's, we might surmise that the floods of 2013 offer us a glimpse into the highly variable weather we might expect in a warmer world. These events certainly got everyone's attention. They clearly demonstrated that our global hydrology is clearly changing. What's more, these changes are beginning to occur faster than our economies and environment can adjust. The loss of hydrological stability is cascading throughout our climate system, undermining the predictability upon which our economy depends for its own stability.

The following year, 2014, the prairies missed a spring flood. Though it was close, so did Alberta. Again the flooding that occurred did not happen in association with the spring melt but as a result of remarkably heavy rainfall in early summer storms. In southwestern Manitoba 50 per cent of the land was covered by water, much of it in the form of floodwaters coming from Saskatchewan. Evidently, what were once loosely considered 1 in 300 year events on the Great Plains have begun to appear every three years.

The prairies are not the only place experiencing this. On July 24, some 25 millimetres of rain – as much as normally falls in a month – fell in 20 minutes on Kamloops, BC. At the time of the storm the entire region was in drought and the area around Kelowna was on fire. Once again we seemed to be witnessing this recurring theme seen also in Colorado of drought ending in flood, often within the same river basins.

On August 4, about 200 millimetres of rain – more than two months' worth – fell on Burlington, Ontario, in the length of time it normally takes to get through the morning rush hour. What were once considered 1 in 50 year floods in this region are now occurring every 10 years. In Toronto and region there have been three 1 in 100 year and six 1 in 50 year storms in the last 25 years. Nor were our neighbours down south spared. As August progressed, flooding of a similar magnitude was experienced in Detroit, Cleveland and Long Island. Another flood occurred in downtown Winnipeg on August 21, and only a week later Steinbach and nearby Niverville in southern Manitoba got hit with up to 120 millimetres – nearly 5 inches of rain – in an overnight storm that overwhelmed flood mitigation systems put in place after flooding in 2002.

Europe went through the same thing in 2014, with catastrophic inundations in England and parts of the continent. As if to round out August in the northern hemisphere, Copenhagen – which had just committed to spending $2-billion over the next 20 years on infrastructure upgrades to prevent flooding – experienced an unprecedented 100 millimetres of rainfall in three hours in its downtown.

COPING WITH INSTABILITY

There is a lot of troubled water out there – and it is trying to tell us something. What it is trying to say is that the hydrological game is clearly changing in fundamental ways. We know, of course, that hydrological conditions on this planet have always been changing, and we now realize how fortunate we have been to have had a century or so of relative stability. That era is over. The long-term hydrologic stability of the climate we experienced in the past will not be returning during the lifetime of anyone alive today. Careful examination of how our hydrology is changing, however, suggests there are ways in which we can adapt.

In a more or less stable hydro-climatic regime, you are playing poker with a deck you know, and you can bet on risk accordingly. The loss of stationarity is like playing with a deck in which new cards you have never seen before keep appearing more and more often, ultimately disrupting your hand to such an extent that the game no longer has coherence or meaning and can no longer be played. This game change means that simply managing land and water in ways that are useful only locally will no longer be enough. We now have to be alert to changes in the larger global hydrological cycle and try where possible to anticipate what they might mean at local and regional levels.

This is a huge new concept – a societal game changer – and it is going to take time to get our heads around it. What we are seeing now is likely what we will be getting in spades in the coming decades, given the most basic of all the laws of atmospheric physics. Predicted rises in temperature of between 2° and 6°C will result in further amplification of the hydrological cycle by 15 per cent to 40 per cent or more. This game change is not going to go away. Until we stabilize the composition of the atmosphere, changes in the Earth's hydrological cycle will continue to accelerate.

It is very difficult to adapt if you don't know what you have to adapt to. Unless we want our future to continue to be a moving target, sooner or later we may have to confront the root cause of our changing hydrology – which unfortunately remains a subject few have wanted to talk about in meaningful terms until recently, at least where I live in Canada.

What has also become clear is that it isn't just emissions cuts that we need in order to restore climatic stability. We can no longer ignore the local value of natural ecosystem processes. In order to gain even partial rein over the hydrological cycle, we have to enlist all the help nature can provide us. We gain that help by protecting and restoring critical aquatic ecosystem function locally by reversing land and soil degradation wherever we can. There is great power in realizing this, for it is at the local level – where we live – that we have the most power to make changes and act most effectively in service of the common good, now and in the future.

We come to the conclusion, finally, that the watershed basin is the minimum scale at which water must managed. This fact in itself – the fact that basin-scale water management is critical to social, economic and environmental resilience in the face of changing hydro-climatic conditions – should inspire our actions. Just like in the movies, this is a time for courageous and relentless citizenship and heroic leadership. The problems hydro-climatic change is bringing in its wake are not going to go away. We should not expect a Hollywood ending. Wishful thinking and last-minute heroics are not likely to get us out of this one. But thoughtful and timely local action will.

CHAPTER TWO

A STORM IS COMING

Emerging Perspectives on the State and Fate of Canada's Water Supply

The writing was already clearly on the wall in 2010. Canada's hydro-climatic regimes were changing, but that was not news the country wanted to hear. In the midst of blistering record high temperatures in the spring of that year, Ottawa was the venue for the final Canadian Foundation for Climate and Atmospheric Sciences conference on the critical role of science in ensuring Canadian water security. Moderated by Gordon McBean, chair of the board of trustees for CFCAS, as it was known, this conference featured presentations of recent findings related to water security in Canada that had emerged over several years from research projects funded by the foundation. The presentations were offered by some of this country's and the world's most respected climate and water scientists.

Opening remarks were offered by Dan Wicklum, who at that time was director-general for water, science and technology at Environment Canada but later went to work for an oil sands industry group. Dr. Wicklum began with first principles. One of the oldest adages in management in general and environmental management in particular, he said, is "if you can't measure it, you can't manage it." In the past, Wicklum pointed out, Canada had a long history of measuring water quantity and quality. In the specific case of managing water and water security, however, simply measuring was not enough. We need to be able to predict supplies and demand, he said, which can be challenging in circumstances where the climate may be changing.

Wicklum stressed that sound policy is not possible without a sound knowledge base. Every step in an adaptive management cycle, he said, from the development and implementation of policy to protect public interest to the tracking of the efficacy of policy directions requires knowledge. Knowledge at each of these steps is largely created by research and monitoring, which is to say that sound policy relies heavily on good science. Dr. Wicklum identified the challenge of conducting ongoing environmental monitoring as being one of the central threats to water security in Canada. His message, which he repeated over and over, was that you can't manage what you don't measure.

There was a certain irony in Dr. Wicklum's observations – an irony which pervaded the entire conference. Wicklum presided over the federal agency responsible for much of the monitoring of water quality and climatic conditions in Canada. His department was in the midst of dramatic funding cuts that would clearly make it far more difficult to manage even what was being measured. Though he may not have intended to do so, Wicklum put into relief what many scientists and policy analysts in the room feared most: that the government of the day did not hold scientific research in the domains of water and climate to be central to its mandate and current priorities. This meant it was likely that the ongoing work of CFCAS would not be supported and that the agency would cease to exist when its current funding ran out.

The opening session was chaired by Elizabeth Dowdeswell, president of the Council of Canadian Academies and later Lieutenant Governor of Ontario, who spoke of the urgent need for relevant science. Dr. Dowdeswell wanted every scientist at the conference to understand that if there was ever a time to demonstrate the value of scientific research to decision-makers, it was now. She succinctly listed crucial unanswered questions in science. Is science generating useful knowledge? Is that knowledge relevant to the problems we face? Is it place-based? Does it empower decision-makers? Is it even available to decision-makers? How do we strengthen and widen the bridge between research outcomes and policy action?

Dowdeswell argued that we need to learn how to create models that encourage researchers to work together and do a better job of bringing social as well as physical science to bear on the central problems of our time. Instead of just organizing more meetings, she urged, science has to generate action.

The keynote address was given by Gordon Young, who at the time was the president of the International Association of Hydrological Sciences. Drawing on decades of international experience, Dr. Young characterized the global freshwater situation and explained why it should concern Canadians. A storm is coming, he said, and we will need all the interdisciplinary science and institutional co-operation we can muster to weather it. Canada, he observed, is not immune to the problems that have created the global water crisis. To avoid crisis, Young maintained, will require a reformed Canadian water ethic.

Dr. Young explained how much fresh water is available to humanity. He illustrated how increases in demand were causing widespread shortages, and in ten minutes he established the clear link between water and food, and water and energy. He observed that while the world was on track to meet the United Nations Millennium Goal of supplying freshwater to the world's unserved, there was little hope of achieving equally important goals with respect to improvements in sanitation. Dr. Young went on to show the direct link between reliable water supply and adequate sanitation and economic and social development.

Dr. Howard Wheater, who in 2010 was the newly appointed Canada Excellence Research Chair in Water Security at the University of Saskatchewan, responded to Dr. Young's address with complementary observations on the growing world water crisis and what it might mean to Canada. Dr. Wheater defined water security as "the sustainable use and protection of water resources, safeguarding access to water functions and services for humans and the environment, and protection against water-related hazards (flood and drought)." Wheater then pointed out that precipitation changes around the world were likely to make water management more difficult in the future. These changes, he said, raise major challenges for science – and for society. He made the point that we need new science, new technologies, new decision support tools and new mechanisms for public engagement to address these challenges.

Dr. Wheater further noted that water managers and policy makers had the same needs as scientists in the context of the planet's changing hydrology. Improved global and regional climate models were needed, and they will have to recognize the limits of predictability. Wheater pointed out that climate models must be improved to ensure they can provide accurate predictions at local scales. There is much to be done to improve Earth-system science too, he said, especially at the level of land surface and climate interaction. It is also very important to assimilate data better, so that adaptation solutions can be robust in the face of considerable uncertainty.

In his conclusion, Dr. Wheater underscored the need to create interdisciplinary collaboration, linking the atmosphere, land and water through integration of natural and engineering sciences with health and social sciences. He added that new science and technologies need to be developed to support integrated water quantity and quality management and to address national and global water security agendas.

In concluding this session, Elizabeth Dowdeswell commented that addressing the global water crisis requires that we harness the intellectual capital and passion of the wider scientific community. We also need to embrace the ethical and moral challenges associated with water issues, and most importantly we need to generate a sense of urgency in addressing water and water-related climate threats.

The rest of the presentations at the conference almost perfectly mirrored not just the content but the urgent tone of the opening session. Hydrologists and atmospheric scientists from across the country offered clear evidence of the fact that Canada's climate is changing and that these changes are being reflected in the timing and type of precipitation and in other factors that could affect water security. Other researchers concentrated on exactly the kinds of measures for addressing these problems that were put forward in the keynote addresses. Not a single speaker denied that our climate is already changing. Each in turn explained why warming temperatures were bound to have a significant influence on the hydrological circumstances of almost every region in the country eventually. Report after report said the same thing. The world our grandchildren will inherit will be very different from the one that exists today.

RECOGNIZING THE PROBLEMS

Among the scientists demonstrating that a storm was indeed brewing were Dr. Ronald Stewart and Dr. David Sauchyn, who summarized recent findings related to climate change effects on the Canadian prairies, especially as they relate to the disturbing prospect of deeper, more persistent drought, not just in that region but throughout southern Canada.

Dr. Stewart, who coordinated the activities of the CFCAS-funded Drought Research Initiative, summarized his network's findings in the following way. The climate forecast for the Canadian prairies suggests there will probably be more drought and at the same time more frequent heavy precipitation events, all signalling the likelihood of greater climate variability. The reason for this is that climate processes appeared to be moving in the direction of greater extremes. A warmer climate will likely accelerate this motion. Why does this matter? Because increased variability in an already highly variable system could have impacts across all sectors of society, which will have major policy implications for all levels of government.

Dr. Sauchyn presented work undertaken principally through the Prairie Adaptation Research Collaborative at the University of Regina. Sauchyn began by pointing out that of the 18 most economically costly natural disasters that have taken place in Canada, 16 were on the prairies. Of those, 11 were droughts. He offered an example: the 2001–2002 prairie drought caused an estimated $2.42-billion loss in crop production in Saskatchewan alone and resulted in a $5.8-billion reduction in the gross domestic product of Canada as a nation.

Quoting from the journal of an early fur trader, Sauchyn observed that major droughts have occurred throughout the historical record and beyond. By carefully analyzing tree rings over a broad area of the Canadian West, Sauchyn and his colleagues have demonstrated that droughts lasting a decade or even longer were not uncommon during the past thousand years. Evidence exists that the wettest two decades on the prairies in recent history were the 20 years when European settlement occurred.

We could not live in such drought-prone places without adaptation in the form of irrigation. Sauchyn noted, however, that there is now increasing pressure in the region to use water resources for purposes other than agriculture. This pressure is mounting at a time when streamflows originating on the eastern slopes of the Rockies are becoming more uncertain.

Sauchyn then introduced findings that showed a strong correlation between drought on the prairies and the 60-year cycle of an ocean-current phenomenon called the Pacific Decadal Oscillation, or PDO. Sauchyn noted, however, that most of our instrumental records – archives of historical data from thermometers, barometers, rain gauges and the like – cover a period of less than one PDO cycle. Climate modelling undertaken by Sauchyn and his colleagues has demonstrated that warmer temperatures will result in a shift toward more positive PDO conditions in the future, which is likely to result in further decline of river flows and even greater climate variability. Extra water will be available in winter and spring, while summers will be drier. Both drought and unusually wet years could occur with greater frequency and severity. This suggests that both floods and deep and persistent droughts are likely to occur more frequently on the prairies, with huge potential costs to the economy.

Dr. Sauchyn also noted that Canada was losing the advantage of cold winters, in terms of both stored water and evaporation. Cold temperatures also created a barrier to disease pathogens that affect both people and agricultural production. He concluded by reiterating that there will greater variation in water and climate in the future, which could have significant impact on agriculture and other sectors of the Canadian economy.

Dr. John Pomeroy then presented new research findings confirming that substantial winter warming is already affecting the hydrology of the mountain West, reducing streamflow and shifting the snowmelt season. Evidence suggests that between 1970 and 2004 there has been a one- to two-month decline in the duration of snowcover in the southern Canadian Rockies.

The conclusions of this research, however, have broad implications that will likely affect every sector of the western Canadian economy. The modelled sensitivity of an alpine snow regime to winter warming suggests that warming of up to 4°C will have a number of effects. It will reduce maximum snow accumulation by up to one-half; reduce seasonal snowfall and sublimation of intercepted snow by more than half; move the date of maximum snow accumulation forward by up to a month; first increase and then reduce snowmelt rates by up to two-thirds; first decrease and then extend the duration of snowmelt by up to one-fifth; and move the date of snowcover depletion forward by up to four weeks.

These results of research conducted by the Pomeroy-directed IP3 (Improving Processes and Parameterization for Prediction in Cold Regions) research network show that intact mountain forests have a mitigating effect on some aspects of climate variability and that windswept open environments are highly sensitive to climate warming. Pomeroy also indicated that mountain forests demonstrably mitigate some aspects of climate variability. As snowpack and forests are clearly going to play a huge role in determining future water security in Canada, Pomeroy urged the research community to press for continuation of the Canadian Foundation for Climate and Atmospheric Sciences so that more could be learned about how to manage landscapes in order to generate optimal water supply in the future.

Environment Canada's Paul Whitfield then introduced research findings from the CFCAS-funded Western Canadian Cryospheric Network (WC²N). The object of this work was to document climate variability and glacier extent in the Pacific Northwest region of Canada over the last 400 years and to predict how glaciers will respond to projected climate change over the next 50–150 years. These research outcomes further confirmed that the hydrology of the West is changing in a direction in which water security will be more difficult to assure.

Through the efforts of IP3 and WC²N researchers, we now know it is possible that we lost 300 glaciers in the Canadian Rockies between 1920 and 2005. Some 150 of those likely disappeared in the 65 years between 1920 and 1985. The other 150 likely disappeared into thin air in the 20 years between 1985 and 2005. This suggests that our losses may be accelerating. The combined findings of the two networks confirmed other studies that had indicated that long before warming reduces the size and output of glaciers in Canada's north and west, it will be changing snowpack and snowcover in ways that could dramatically affect water supply.

Reduced flows in Western rivers, Whitfield noted, will impact power generation and agricultural, industrial and municipal water security throughout western Canada. These flow reductions will also affect interprovincial water sharing arrangements and transboundary agreements with the United States such as the Columbia River Treaty.

Hydrologist Garth van der Kamp then shared the results of his recent research on the role groundwater will play in defining water security in Canada. Many in the audience were surprised by his findings. Dr. van der Kamp offered evidence that prairie groundwater is not recharged from the Rocky Mountains as many believed; rather, nearly all the groundwater in Canada is found in small aquifers. This suggests that in the event of need, these aquifers can only supply large amounts of water for short periods. Water security over time will require careful management of groundwater to prevent overdraft.

At present, van der Kamp explained, the security of groundwater is being widely diminished by contamination. The fact that one in four Canadians rely on groundwater for domestic supply and a million people rely on wells that already exceed federal contamination guidelines suggested to van der Kamp the need for co-location of ongoing research in experimental watersheds.

Dr. van der Kamp's research demonstrated that groundwater can provide water security during droughts – provided it is left in reserve and not overexploited. At present, however, we continue to separate groundwater and surface water in our management approaches, ignoring the air–soil–groundwater–surface water interface that is the very essence of how the hydrological cycle works, especially on the Canadian prairies. Monitoring groundwater, van der Kamp concluded, should be a priority, because without clear knowledge of what is happening to it, adaptation to climate change may become very difficult. "We have to realize the seriousness of the problem," van der Kamp said quietly, "and then act on it."

Philip Marsh then moved the focus of the conference to the North and one of the largest features of its kind in the world, the 12,000 square kilometre Mackenzie Delta. Dr. Marsh explained the influence of the flows of the Mackenzie on the 40,000 lakes found in the region and the delta's influence on the Beaufort Sea and its ecology. Dr. Marsh and his colleagues have discovered that the delta is subsiding, causing water levels to change. This subsidence will affect ecological function as well as actual and projected gas pipeline operation. Marsh argued that an improved knowledge of the effect of the Mackenzie Delta on flows of water, sediment and nutrients is very important to understanding the Beaufort Sea, especially considering the likelihood of future changes in the Beaufort's ecology.

Marsh observed that over the last 15 years, Canada has made significant investments to improve our science capacity in order to both understand and improve our ability to predict the weather, climate and hydrology of the Mackenzie basin and delta. These programs include the Beaufort and Arctic Storms Experiment; the Mackenzie Global Energy and Water Experiment; the CFCAS-funded IP3 initiative; and Canada's Northern Energy Studies pertaining to the Mackenzie gas project. These research undertakings were all advanced in one way or another through the International Polar Year, which focused in part on changes in Arctic river deltas.

Dr. Marsh lamented that just when much more can be known as a result of the recent development of coupled climate models, research is ending. He concluded by explaining that what we presently need most is follow-up support to allow the newly created models to use data collected in these research projects to generate policy-relevant outcomes.

The next presenter, Sean Carey, showed results which confirmed that climate change effects are already being dramatically felt by way of widespread landform changes wrought by permafrost melt in the Arctic. Change in the North is now the norm, not the exception. Nowhere on Earth, Carey observed, is the landscape changing more quickly. The rate of change in the Arctic has been much more rapid to date than all but the most extreme projections, Carey said. And most of the changes involve water.

Carey and his IP3 network colleagues were working to determine how a changing climate and accelerated human activity will affect the northern water cycle. The network, which functioned from 2006 until 2010, was composed of about 40 investigators and collaborators from Canada, the US, the UK and Germany. Carey explained that IP3 was devoted to understanding water supply and weather systems in cold regions at high altitudes and high latitudes, which explains the focus on research in the Rockies and the western Arctic. The network has already contributed to better prediction of regional and local weather, climate and water resources in cold regions, Carey said, including understanding of streamflow in places where there are no gauges. The research has also led to advances in understanding of changes in snow depth and cover and the resulting effects on water supplies and calculation of freshwater inputs to the Arctic Ocean.

Carey went on to point out why the hydrology of the Arctic is so sensitive to upward changes in temperature. Permafrost acts as an aquitard – a formation through which water flows transversally much slower than it does through an aquifer, and which also restricts vertical movement because of the impervious nature of subsurface ice layers. In losing permafrost, the landscape is not only losing the rigidity provided by subsurface ice, but water is moving downward in ways that have not been witnessed before. Everything with respect to hydrology appears to be changing at once.

The Fort Simpson area of the Northwest Territories, Carey indicated, has experienced a 30 per cent decrease in permafrost in the last 53 years. Ponds are shrinking in the Arctic as a result of internal drainage made possible by the degradation of shallow permafrost. Streamflow chemistry and water quality will also be impacted as the movement of water in the subsurface changes. Changing climate also affects the timing and magnitude of floods, breakup and river freezing, which is already causing considerable damage to northern infrastructure such as culverts, dams, bridges and roads. Carey noted that at present we have a poor scientific capacity to predict flow – or floods – in small and medium-sized northern streams.

Carey noted that road and pipeline river crossings in the north routinely fail because we apply insufficient understanding of hydrology, runoff generation and permafrost or frozen ground dynamics in our designs. Carey also indicated that across the northwest of our continent, certain tree species no longer respond to climate signals as they have in the past half-millennium. There is also evidence that ecological thresholds have been reached. Like many of the other researchers at the conference, Sean Carey concluded by underscoring the importance of ongoing research in this area to the building of further adaptive capacity in both northern and southern Canada.

Next, Dr. John Hanesiak of the Storm Studies in the Arctic Research Network shared groundbreaking new results tying higher temperatures to changing precipitation and storm patterns in the Arctic. Storms and their related hazards in this region have profound effects, including loss of life and impacts on industry, transportation, hunting, recreation and even the landscape itself. Such storms also impact sea ice and marine systems. Hanesiak demonstrated that over recent decades the frequency and intensity of storms in the Arctic have been increasing.

In the Arctic, ecosystems and people can be dramatically affected not just by prolonged blizzards but also by windstorms and unusual warming events. In 2006, for example, February temperatures of +5°C, as opposed to the normal seasonal values of around -20°C, brought about by an intense cyclone across southern Baffin Island, caused rain showers, while accompanying 125 kilometre an hour winds caused property damage and made travel impossible. In June of 2008, some 35 millimetres of rain fell in 12 hours at Pangnirtung, Nunavut. Temperatures soared to more than +13°C, with winds of 80 kilometres per hour. Significant snowmelt followed, causing a stream to take a new and deeper course through the community. The town remained physically divided until new bridges could be built.

In combination with permafrost loss, extreme weather events are expected to play havoc with northern development. Unfortunately, however, the lack of observational sites makes it difficult to know exactly what is happening in the Arctic. The surface and satellite-based research Dr. Hanesiak and his colleagues were undertaking was aimed at improving understanding and prediction of severe Arctic storms and their hazards. The team's models demonstrate that in future there are likely to be more days with precipitation at northern latitudes and more rain days, which will likely result in less ice. The models also projected an 8 per cent increase in freezing precipitation per decade, which could have a significant impact on herd species such as caribou and muskox. This is bad news given the already diminished state of many of the largest caribou herds currently in the Canadian North. The potential effect of more frequent and intense storms on potential hydropower developments and other infrastructure is unknown.

One of Canada's most respected scientists, Jim Bruce, then told the audience that IPCC climate change projections are likely too cautious and that some effects are already occurring faster than expected. Despite this, he said, emissions continue to rise, with the global total in 2010 topping the 1990 level by 41 per cent. As a result, impacts are becoming obvious: glaciers are visibly receding; ice on Canadian lakes and rivers is disappearing; and extreme weather events are becoming more common. Dr. Bruce observed that if the changes that have taken place over the past 40 years continue into the future, Canada should expect to experience significant climate-related impacts on its water supplies. He concluded by pointing out that in the Canadian context, climate security equals water security. Without the one, you don't have the other.

It's not just the Arctic that is changing, either. Research outcomes of projects supported by the CFCAS also demonstrated how climate change was beginning to alter ecosystem function in some of the most heavily populated areas of the country. Linda Mortsch, for example, explored impacts on water and aquatic ecosystems in the Great Lakes region, notably the interconnected effects of lower water on habitat suitability for wetland vegetation, birds and fish. Research at Long Point on Lake Erie demonstrated that a 1.48 metre decline in water level resulted in a 32 per cent decrease of the surface area of the inner bay. The shoreline grew from 200 metres in width to more than 2 kilometres. Mortsch noted that water level variability maintains vegetation, bird and fish habitat, and that changes in this variability as a result of climate will have implications for biodiversity and ecological services.

Turning attention to the prairies, Dr. Mortsch observed that "prairie potholes" are particularly sensitive to water balance changes. In drought conditions, permanent potholes become ephemeral, altering wetland vegetation and wetland functions and values, with significant effect on waterfowl habitat. Mortsch further observed that during drought such wetlands become vulnerable to being converted into agricultural land.

Her team also identified the effects of warming water temperatures on aquatic ecosystems, with particular reference to changes in water chemistry and the effect of that on algae populations and invasive species. All of this, Dr. Mortsch pointed out, would have important implications for water security in the future.

IDENTIFYING THE SOLUTIONS

While the research outcomes of the physical scientists clearly revealed growing widespread threats to water security as a result of current management policies, which will be exacerbated by climate change, policy analysts who spoke at the conference demonstrated that solutions exist and are still within our means. In fact, important steps forward are already being made. The challenge always is to keep up with the problems we are creating for ourselves under current water governance regimes and under the weak climate change policy formally adopted by the former Harper government.

In order to put into relief how we arrived at where we are in Canada with respect to water and climate security issues, water policy scholar Ralph Pentland tracked and compared the role of science in decision making and policy development over four 20-year periods in the history of Canada–US water relations. His research revealed that the gradual hollowing-out of government itself and the decline of government science over time have eroded the sustainable management of water resources in favour of short-term economic interests. This, Mr. Pentland pointed out, has adversely affected relations between the two countries, as exemplified by the return of water quality problems in the Great Lakes and growing difficulty with algae blooms in Lake Winnipeg.

Pentland observed that between 1965 and about 1990, there was a dramatic change in conventional wisdom about governance. We suddenly realized, he noted, that one person's effluent was another person's intake, and that the public good did not always coincide with the pursuit of private interest. The high degree of interdependence created by technology suggested the need for a total "systems" approach in the search for optimal societal solutions, especially as they relate to the environment. Pentland observed that as we moved into the most recent period after about 1990, globalism and competitiveness agendas began dominating conventional wisdom and governance. Even though we claimed to be operating under the banner of sustainable development, the agenda had become the so-called virtuous circle – the assumption that global economic growth, the promotion of democratic systems and the encouragement of international trade and investment would automatically produce a self-reinforcing cycle of wealth generation, social advances and eventually ecological protection.

The conventional wisdom about governance will have to change, along with the perceived role of science in decision-making, Pentland predicted, if only because the present system is not sustainable and will be even less so in a world in which the climate is changing. Real ingenuity will be required, he said, as Canada as a nation is forced to work its way back to sustainability. We will have to catch up with our American neighbours in the areas of monitoring and pollution control, Pentland concluded, if we don't want our relations with the US over water resources to deteriorate.

In a similar vein, Ted Yuzyk of the International Joint Commission (IJC) talked about the importance of enhancing Canada's scientific capacity with respect to the joint management of transboundary waters. He explained that all of the IJC's boards relied heavily on sound science, particularly in the form of comprehensive and standardized data. Mr. Yuzyk lamented the absence of long-term standardized monitoring information.

Noting that water covers 43 per cent of Canada's 8900-kilometre border with the United States, Mr. Yuzyk identified a broad range of transboundary issues that need to be considered jointly by both countries. These included climate change and variability, deterioration of water quality, the appearance of invasive species, water apportionment, and joint response to floods and droughts. Fundamental to addressing these issues, he reiterated, is the need for comprehensive, standardized data. We need to upgrade our models and modelling capacity with respect to changing hydrology, surface and groundwater quality and change in climate, he said.

It was interesting that the remarks of Yuzik, as one of the last speakers in the conference, mirrored those of the very first presenter: if you can't measure, it is hard to manage. Yuzyk also observed, reflecting on what Ralph Pentland had said, that in the 1990s Canada had gone the wrong direction with its water monitoring strategies. Both nations now desperately need sustained, harmonized hydrologic data, Yuzyk said, and without it there would be no assurance of good relations with our American neighbours.

Al Pietroniro of Environment Canada's National Water Survey demonstrated that real progress is in fact being made on water security problems in Canada. He began by illustrating innovative new ways of coupling inputs of research data that could vastly improve the predictive capacity of weather and climate models. Linking Environment Canada's MESH hydrology model with the Canadian Precipitation Analysis model, for example, combines different sources of information on precipitation into a single, nearly real-time analysis. Analysis is then used to improve environmental predictions, which can now include satellite observations of soil moisture.

Pietroniro went on to share progress in modelling and parameterization that had emerged from the CFCAS-funded IP3 research network. He noted that innovation and networking have allowed for measurable advances in Environment Canada's water cycle modelling system. Coupling atmospheric and hydrological systems allows for advances that are of value to the entire water resources management community. A consistent and coherent "community approach" to modelling has resulted in systematic improvements in the modelling framework. These improvements include greater efficiencies in software development and maintenance and growing rigour in the design and development of models. These innovations can and do make their way directly into the operational systems decision-makers rely on for information, which underscores the value of partnerships between universities and governments in water and climate research. The further development of these models, Pietroniro indicated, would be central to the enhancement of Canadian understanding of – and adaptation to – future climate circumstances. It was clear, however, that the results of these technological innovations need to find their way into the decision-making process more quickly than they do today, which was the subject of the final presentations at the conference.

Slobodan Simonović offered a new way of thinking about water management in an era of climate change in central Canada. He was unequivocal in stating that climate change is real and more serious than expected. Dr. Simonović made the point that water is the delivery mechanism for many of the impacts of climate change, and that water-related feedbacks will likely be a central feature of climate change effects over much of Canada. Global warming, he agreed, will likely enhance both extremes of the water cycle. We should expect more intense droughts, heat waves and wildfires. We should expect heavier rainfall, more floods and stronger storms. Adaptation to these will demand new ways of managing water based on emerging understanding of systems dynamics.

Our current management traditions, Simonović observed, attempt to manage the environment by addressing water, land and air issues separately. Whenever we push at one point in the environment, however, it results in unexpected changes elsewhere. Dr. Simonović pointed out that this was in fact one of the fundamental features of an integrated system. This realization suggests there is a need for a broader, "systems-based" view of management options.

A new way of viewing the problem involves accepting that *interaction* between socio-economic and natural systems causes climate change. Interaction determines the entire system's evolution. The systems approach establishes the proper order of inquiry and helps in the selection of the best course of action that will accomplish a prescribed goal. By broadening the information base of the decision-maker and providing a better understanding of the system and the interrelatedness of its component subsystems, systems analysis facilitates prediction of the consequences of several alternative courses of action.

Dr. Simonović proposed putting hydrology at the centre of a new decision-making model where concerns related to water radiate outward to inform the way we function as a society. The creation of such a decision-making model, he concluded, would rely on science's ability to identify the tipping points that will inspire politicians to act. Like all the other speakers, Simonović had only twenty minutes. But the import of what he said radiated throughout the room. What he proposed was nothing less than turning the decision-making world on its head and putting water back into the position it once occupied of being central to who we are as a people and what we mean as a nation.

The next speaker, Dr. Hank Venema, presented findings that linked crisis and vision to future opportunity. He observed that there are presently a number of gaps in water management as it is presently practised in Canada and abroad. The water management sector, he argued, urgently requires new institutional capacity that can merge the natural sciences, the social sciences and public health concerns with engineering and public policy innovation so as to enable us to overcome the jurisdictional fragmentation that currently characterizes water governance. He also noted that the sector requires a much greater use of innovative financial and investment instruments that reinforce the values of integrated water resources management and restore natural capital.

Dr. Venema proposed that one solution would be to recast our current "commingled crises" as an opportunity to integrate climate change concerns, hydrology and water quality issues and nutrient scarcity into new economic possibilities. Venema noted that adaptation will require corporate and community collaboration at the intersection of social and technological innovation, and he illustrated how these ideas could be applied to problems associated with nutrient loading in Lake Winnipeg, currently the most eutrophic large lake in the world.

Extended forms of collaboration, Dr. Venema concluded, will have to become the norm in Canada if we are to address serious and growing concerns related to water and climate security. We need to respond early to the storm warning, he said, so that we can minimize its impact on our future.

SEEING OUR WAY THROUGH DARK CLOUDS

Five points appeared consistently throughout the broad range of research outcomes and related presentations at this Ottawa conference. The first was that we are already seeing significant changes in the climate and hydrology of the country. The evidence is obvious and visible. Changes have been measured and observed everywhere in Canada. All the research presented suggests this could be a problem for, if not a threat to, water security.

Climate change effects largely relate to water. The problems are already serious in some areas and could have huge consequences economically, socially and politically in the future. Deep and persistent drought could devastate our agricultural sector. Groundwater overdraft will diminish our adaptability to drought. Long before warming has diminished the size of our glaciers, it will begin to reduce snowpack and snowcover, which could dramatically and directly affect water security. The Arctic as we knew it is vanishing right before our eyes. And yet – perhaps because scientists are trained to be objective in such matters – researchers act as though they are resigned to these outcomes. This resignation only serves to preserve the status quo, especially as it relates to political inaction on climate threats to water security.

It became clear through the conference proceedings that we did not have any real idea what we were signing up for when, as a nation, we decided it would be cheaper and less inconvenient to adapt to climate change rather than address it. By choosing adaptation without thinking about the consequences of the climate threat, we have been forced, without even discussing the matter, to agree that we won't mind dealing somehow with hordes of environmental refugees; conflicts locally and abroad over water supplies; the enormous costs of protecting coastal cities from rising sea levels; and huge outlays for repairing damage to our cities caused by more frequent and intense rain, snow and ice storms and floods.

By choosing to adapt rather than act on the climate change problem, we are agreeing that we accept the loss of the Canadian Arctic as we know it and that we will somehow manage through 45°C summer temperatures on the prairies and absorb the enormous economic costs associated with deep and prolonged drought that could lead to food shortages. Because we are choosing to adapt rather than respond, we have accepted that we will live with more frequent wildfire, an increase in invasive plant and animal species, and the likelihood of new diseases, all of which will result in a lower quality of life for ourselves and our children.

The second important realization to emerge from this conference was that just when we are beginning to find out what we really need to know about climate effects on water security in Canada, we are terminating critical scientific projects. The findings from research networks funded by the Canadian Foundation for Climate and Atmospheric Sciences were of national if not global significance. Over the last seven years researchers had planted a spectacular garden and now they were forced to sit by and watch as it got hailed out. The subject of disasters surfaced several times during the conference. The termination of support for the CFCAS was itself a disaster.

Another thread running through these presentations was that there is a great need in this country for more interdisciplinary and interinstitutional co-operation on research related to water. We know what we need to do to create an interdisciplinary domain, but we have to do it. As Slobodan Simonović pointed out, water is the delivery mechanism for many climate change effects. We need to use water science as a vehicle for better land-use planning and more effective management of our water resources. Better understanding of our water resources may be the cheapest, simplest and most direct way to adapt to climate change.

The fourth theme to become apparent from this conference was that there is an urgent and ongoing need for federal institutions like the CFCAS to fund network research. Individual scientists cannot, and may not wish to, concern themselves with linking research outcomes to public policy action. But backed by research networks, they can. If nothing else, presentations at this conference suggested that more of the budget of any given project has to be committed to making research outcomes intelligible to others. Networks have to be far more directed and forceful in getting their results onto decision-makers' agendas. One wonders if this fact doesn't put into relief the need to create a whole new interdisciplinary field that could build a far more effective and durable bridge between science and public understanding that ultimately, through policy action, would lead to greater water security in Canada.

A related idea emerging from these sessions was that there is also a need to preserve an existing bridge between science, public understanding and public policy action on climate-related threats to Canadian water security. This crucial need to connect research outcomes with public policy choices was affirmed once more in the final keynote presentation. This address was offered by the only sitting politician to accept an invitation to attend the conference.

The Hon. Michael Miltenberger, at the time deputy premier and minister of environment and natural resources of the Northwest Territories, spoke about politics and water. He began by outlining the details of the Northwest Territories' new, jointly developed *Northern Voices, Northern Waters* strategy. He observed that in the context of implementing this groundbreaking water policy, the building up and passing on of knowledge about climate and water issues in the Northwest Territories were of crucial importance to his government. In this, he explained, science was not enough. He noted that politicians could not afford the luxury of the singular intellectual focus enjoyed by scientists. He went on to indicate that politicians like himself needed scientists and vice versa.

Miltenberger urged conference participants to find ways to break down silos in both science and government in Canada so as to arrive at a common vision and set of principles regarding the management of our country's water resources. Everything he had learned as minister, Miltenberger said, had told him that Canada needed a nationwide water strategy and that water security demanded the creation of such a strategy.

In closing it was held that this final Ottawa conference could be a turning point in the history of Canadian science. With the passing of the Canadian Foundation for Climate and Atmospheric Sciences we may very well ask ourselves the following questions. Are we going to gravitate to the lowest common denominator in our response to climate change effects on our country's water resources or are we going to take charge of the problem? Are we going to be satisfied with doing less with less or are we going to properly support research that will yield answers to questions we desperately need to ask about our future as a nation? And are we going to take those answers seriously?

It may be time to recognize the urgency of these questions. The failure to do so – as was pointed out many times at this important conference – may result in our losing our way as a country and causing the loss of our place in the world. There was not a person present who wasn't dedicated to preventing that from happening.

One might wonder how far south from the Arctic, how far downslope from mountains and how far inland from the seas the effects of climate change are going to have to advance before we get serious about its potential effect on the nation's water security. The hydro-climatic time bomb we have created is now ticking loudest in the Arctic, where the landscapes that people historically held to comprise their country have begun to melt out from under their feet. It is in the Arctic that the story of water and climate security really begins and ultimately may end.

CHAPTER THREE

ARCTIC TRADE-OFFS

The Implications of a Rapidly Warming North

The global atmosphere is an ever-changing reservoir where what enters the air by way of marine and terrestrial sources is subjected to a broad range of oxidation processes before being recycled into the world's oceans or back to the land surface. The two main mechanisms by which this is accomplished are through erosion and wind transport and through the scavenging of these chemicals in the atmosphere by rain and snow and their return to the Earth and its oceans by way of precipitation. Rain and snow carry a great deal of planetary freight. As it happens, snow is a better scavenger of particulates than rain is. This is a significant factor in regions like the Canadian Arctic that are covered by seasonal snow for much of the year.

When we think of Canada's cold regions, obviously we think of snow. But snow is never just snow. Falling snow carries within it small bits of the Earth's crust, usually in the form of dust composed principally of calcium or magnesium. It also carries weak acids and trace metals along with neutral organic materials such as pollen, airborne plant parts, seeds and insects and other living things small enough to be carried up into the atmosphere by wind. Like liquid water, snow also carries within it almost everything humans dump into the atmosphere, including contaminants like polychlorinated biphenyls (PCBs), persistent organic pollutants (POPs) and mercury, all of which are introduced into Arctic ecosystems from the south. Many of the most problematic pollutants are carried northward by globally circulating winds and brought to the ground in the snow that falls in the Arctic.

The concentration of atmospheric pollutants in the Arctic is not an easy problem to address, for it is planetary in origin. As our world spins beneath the thin clothes of its atmosphere its very rotation acts rather like a centrifuge, drawing from the equator to the poles everything put into the air anywhere in the world. These airborne contaminants are brought back down to Earth as a result of being scavenged from the atmosphere in the process of snow formation and the merging of multiple crystals to form larger snowflakes. These substances find their way into snow at the point of its formation. Ice crystals most often form in the atmosphere as a result of the freezing of super-cooled water droplets at -40°C, a temperature not uncommon during the Arctic winter. Researchers have determined that ice crystal formation is catalyzed by a process called contact nucleation. Ice nuclei normally consist of sea-salt aerosols, particulate organic debris or fine particulate clays that originate as dust motes. Ice crystals may form by direct deposition of water vapour onto ice nuclei or by water droplets either touching ice nuclei or through the scavenging of nuclei.

It is difficult to imagine just how much material can be transported in the atmosphere and deposited by way of snowfall. Even though much of the high Arctic is a cold desert, the accumulation of atmospheric contaminants from the south in the cold winter air turns snow into a precipitate of diluted poison. Wind gathers the contaminants in drifts, where they wait to be mobilized by the spring melt. Contaminants are further concentrated when the snowdrifts feed spring flows. In the meantime, however, these contaminants bioaccumulate in the bodies of the animals that drink the water and in the bodies of the people who depend on these animals as their food source.

Few Canadians appear to be aware of the fact that economists around the world are now warning of an emerging new scarcity that will affect every sector of the global economy. This growing penury is defined by very real and increasingly apparent limits to the capacity of the environment and our water to absorb and neutralize the unprecedented streams of waste that humanity is releasing into it. The gravity of this problem may not become obvious until we realize that we ourselves are becoming walking contaminant sites. Nowhere is this more evident than in the Arctic. In a recent study, Canadians were tested for the presence of 88 harmful chemicals in their bodies. A First Nations volunteer from a remote community in Hudson Bay was found to have 51 of the chemicals in her system. Arctic environments are so contaminated with substances that have been carried north by the global atmosphere and deposited there by rain and snow that some Inuit women have been advised to begin weaning their babies earlier in order to reduce exposure to contaminated breast milk.

THE ENVIRONMENTAL CHALLENGES

AIRBORNE MINE TAILINGS

While oil sands operations in Alberta are hardly the only source of airborne particulate matter in Canada, they demonstrate the nature of the problem. A December 2009 paper in the prestigious US journal *Proceedings of the National Academy of Sciences* reported that oil sands development in the Athabasca River basin is a greater source of contamination than previously realized.

Lead author Erin Kelly, a post-doctoral researcher working under the supervision of Dr. David Schindler at the University of Alberta, had discovered that during four months in the winter of 2008, some 11,400 tonnes of airborne particulates fell onto the snowpack within a 50 kilometre radius of the Suncor and Syncrude bitumen upgrading facilities. Of this amount, an estimated 391 kilograms were found to be polycyclic aromatic hydrocarbon compounds that are known to be toxic to aquatic life.

This is equivalent to having 600 tonnes of bitumen fall from the sky. At some of the monitoring sites, an oil slick had formed on the surface of the melted snow. Some sites showed that the concentration of polycyclic aromatic compounds present in tributary streams of the Athabasca increased by as much as eight times normal as a result of nearby mining and related activities, essentially rebutting the argument that all of the contamination in the area's watercourses is an entirely natural consequence of water streaming through exposed bitumen deposits. The report indicated that over the course of a full year an estimated 34,000 tonnes of particulates falls annually within that 50 kilometre radius, or nearly five times the volume of emissions reported by the industry. This, the report claimed, was the equivalent of a major oil spill each year.

Since particulate deposition rates in the 1970s were as great as they are today, because smokestack precipitators had yet to be installed, this situation has likely persisted for 30 to 40 years. As a result, current background concentrations of polycyclic aromatic compounds in surface soils, vegetation, snow and runoff over a broad area of the boreal forest may be measurably greater than true background concentrations contributed naturally by oil sands deposits in the region. Fortunately, the effects of these particulates appear to decline by the time the waters reach Lake Athabasca, though deformations in fish could still result from contaminant levels of this magnitude upstream.

Kelly's PNAS paper confirmed the serious defects of the joint government, industry and stakeholder operated Regional Aquatic Monitoring Program, which it criticized for lacking scientific oversight and peer review of monitoring and research results; failing to provide data to the public; and using faulty methods to analyze, interpret and report data in a timely and transparent way. The paper argued that more than 10 years of inconsistent sampling design, inadequate statistical power and the use of "monitoring-insensitive analytical practices" had resulted in the failure to detect major sources of toxic contaminants in the Athabasca watershed.

In conclusion the Kelly paper recommended oversight by an independent board of experts to make better use of monitoring resources and ensure that data were available for independent scrutiny and analysis. In order to limit bad press abroad, both federal and provincial governments scrambled to respond to the need for improved independent monitoring, which has since been established. Though promising, it remains to be seen whether renewed efforts to accurately measure the airborne effects of such oil sands operations will change anything. It is clear, however, that the problem of contamination of our snowfalls is becoming serious, especially in the North. The scale and intensity of oil sands development and the complexity of the manner in which polycyclic aromatic compounds are transported in the atmosphere demand no less than the highest quality of scientific attention. The fate of the Athabasca and of the entire Mackenzie River system of which it is part depends on it.

PERMAFROST LOSS

Permafrost is defined as ground that remains at a temperature below 0°C for at least two consecutive years. Permafrost can take a number of forms. It can comprise bedrock, sediment, soils or organic materials such as peat that may or may not contain ground ice. It has been estimated that permafrost in one form or another lies beneath more than 22 million square kilometres of the Arctic and sub-Arctic regions, and that some two million square kilometres – nearly 10 per cent of the Arctic region – is underlain by ice-rich permafrost.

Though all permafrost can be altered by warming and thawing, the melting of ice-rich permafrost is particularly problematic in that it results in feedbacks that produce noticeable effects on ground surface stability, microtopography, hydrology, ecosystem function and the carbon cycle. A recent monitoring initiative drilled some 850 boreholes in 200 active permafrost layers throughout the Arctic. The results demonstrated a general warming trend in permafrost layers throughout the 1990s and early 2000s. Evidence suggests that warming has been most pronounced in the northern continuous permafrost zone and most recently along the Arctic coast. Canadian researchers, however, have also discerned rapid permafrost loss in the discontinuous permafrost zone in the southern regions of the Northwest Territories. This observation is consistent with findings in Russia which suggest that while warming in the continuous permafrost zone has generally not resulted in thawing during the past 60 years, widespread thawing in the discontinuous zone appears to be occurring throughout the northern circumpolar world.

Researchers have also discovered that there is a great deal we don't know about the dynamics of permafrost thawing. It appears that vulnerability of permafrost to climate warming is complicated by surface properties such as snowcover, vegetation, the thickness of the topmost active layer of the soil above the permafrost, which freezes and thaws seasonally, and interactions between surface water, groundwater and soils. Changes in these complicated relationships can lead to both positive and negative feedbacks to permafrost stability, allowing permafrost to exist in some areas at mean air temperatures as high as +2°C along the southern margin of the discontinuous permafrost region and at the same time degrade at mean temperatures as low as -20°C in the high Arctic.

Continuous vegetation cover can also reduce mean soil temperatures by as much as 6°C. In contrast, the presence of standing surface water can raise temperatures at the water–sediment interface by as much as 10°C, which can make permafrost vulnerable to thawing at even the highest northern latitudes. While resulting changes in landforms due to permafrost thawing are of growing concern everywhere in the Arctic, the threat of loss of terrain stability is only one problem this thawing is causing.

PERMAFROST CARBON EMISSIONS

Dr. Katey Walter Anthony is an associate professor at the University of Alaska in Fairbanks who is an expert on what is currently known about the critical relationship between permafrost loss and methane and carbon releases in the Arctic. Permafrost, Walter notes, contains approximately 1700 gigatonnes of carbon, or roughly twice the amount that is presently found in the atmosphere. Release of this carbon from thawing permafrost is a well-recognized global risk because it may cause feedback amplification of the anthropogenic warming due to greenhouse gas emissions already occurring elsewhere. This feedback begins with the thawing of organic matter containing carbon dioxide and methane, which in itself increases the surface temperature, further warming the top layer of frozen soil, which in turn accelerates further melt and further carbon dioxide and methane release.

Walter explains that there are two permafrost observation networks in the world, and both were asking the same two questions: How much methane and carbon dioxide is being released through permafrost thaw? and How important will permafrost carbon feedback be in a future warmer world?

To begin answering these questions, Walter investigated the gas bubbles which rise from thermokarst lakes in northern Alaska and Siberia. (Thermokarst terrain is land dented with depressions caused by soil collapse when the underlying permafrost thaws. Thermokarst lakes, also called thaw lakes, form when meltwater gathers in these depressions.) In summer, methane bubbles up through the water in these lakes and is released into the atmosphere. In winter, however, the methane collects under the ice and, if suddenly released, can be ignited. There are millions of lakes in the Arctic accumulating and bubbling gases. What we need to know, observed Walter, is how much of an issue this might be in terms of future climate effects.

Walter and her colleagues have surveyed some 200 lakes in the northern hemisphere and have discovered that they can be classified into "yedoma" and "non-yedoma" lakes. (Named for the Yedoma region in Siberia, yedoma is a particular type of extremely ice-rich permafrost.) It is the yedoma lakes that matter. This type represents only 9 per cent of the region's lakes, but they are a source of 35 per cent of the methane such lakes produce. Yedoma lakes are Pleistocene in origin. The soils that surround them are composed of between 50 per cent and 90 per cent soil carbon. What is interesting about these soils is that decomposition had not advanced far before permafrost locked them in cold.

By studying the yedoma soil permafrost profiles, Walter calculated that some 14–34 teragrams of methane was being released each year in the Arctic from lakes and that total carbon dioxide and methane releases from ecological sources were in the order of 50 teragrams a year. Walter and her colleagues projected that Arctic marine systems were likely contributing an additional 1–17 teragrams of methane released per year. Because the Walter team were unable to determine the extent of release of methane from geological sources such as pockets in subsurface geological strata, the total methane and carbon dioxide emissions from the northern circumpolar boreal region remain unknown.

Walter has, however, been able to determine how current permafrost thaw and resulting greenhouse emissions might compare with releases in the past. She did this by examining sediments that show the progression of change over time during the life of a typical northern thermokarst lake. The end point of this change appears to result in lakes eventually being drained, at which time the lake bottoms typically refreeze. Walter then examined how much carbon and methane was stored in Holocene sediments.

Research into this topic revealed some surprises. Walter and her colleagues observed that yedoma lakes are sequestering carbon at a rate five times faster than any other lake system. The reason for this is that these lakes are cold and deep, and when trees or other organic materials fall into them, decomposition is very slow.

Walter developed a time profile showing when the yedoma lakes were formed and then laid that time profile over a profile of associated methane releases and carbon sequestration. She then laid different radiative forcing scenarios over the same time profile. These profiles in juxtaposition showed the cumulative change in time of the permafrost soil carbon pool. Walter concluded from these analyses that permafrost thaw beneath yedoma lakes was an important source of atmospheric methane and a positive feedback to climate warming in the early Holocene. She also noted that permafrost thaw also contributed at the same time to carbon sequestration. The most important finding for us today, however, is that future widespread permafrost thaw will reverse the mitigating role of Arctic lakes by releasing into the atmosphere much of the carbon sequestered in these systems for the past 10,000 years.

The next question Walter addressed was whether or not methane hydrates could be expected to threaten humanity with further climate disruption in the future. Yes, Walter said, methane hydrates are a concern but not immediately. Present-day emissions from hydrates are low. Walter and her colleagues estimated that centuries of warming would be needed to induce large-scale hydrate releases from permafrost. That said, Walter also noted that she and her colleagues were witnessing very large lake seeps, which showed there was a big difference between ecological sources of methane release, and geological seeps, which can be a constant source of methane bubbles. The number of geological seeps is potentially very large, Walter reported, perhaps as much as seven orders of magnitude more numerous than ecological seeps. Ecological seeps, she noted, are bacterial in origin, while geological seeps are far more complex and most often related to the proximity of coal formations and natural gas reservoirs near or linked to the surface.

What we have, Walter concluded, is "a very leaky cryospheric cap" and this cap is deteriorating under warming conditions. At present, permafrost caps methane and carbon dioxide releases, but the cap is melting and glaciers are disappearing, causing decompression of subsurface geological structures and isostatic rebound, which releases thermogenic methane. Similar processes, Walter noted, are clearly active in areas where glaciers are disappearing in Greenland and Iceland.

Returning to the question of whether the catastrophic release of methane hydrates in the circumpolar Arctic should trouble anybody's sleep, Walter noted that the model projections of cumulative emissions from methane releases suggest that permafrost carbon feedbacks could be very important as global temperatures rise in the future. Assuming that biogenic methane represents 2.3 per cent of the carbon emitted from thawing permafrost, Walter and her colleagues estimated that this source could represent 16 per cent to 50 per cent of warming due to permafrost carbon feedback by 2100. This amount represents 1 per cent to 3 per cent of the warming projected to be due to anthropogenic emissions. This suggests a global temperature increase of 0.29°C from permafrost carbon feedback.

Walter noted also that, at present, permafrost carbon feedback is not included in future warming scenarios put forward by the Intergovernmental Panel on Climate Change. Carbon decay of organic materials takes 150 years, she observed, which means that permafrost thaw will release substantial amounts of carbon after 2100.

Walter has also extended her analysis to the potential commercial value of both biogenic and geological releases of methane. She noted that in discussion with energy companies these releases were evidently not considered to be of economic value, because they were too dispersed.

Also put forward, in the context of the rate of deterioration of the cryospheric cap holding back the vast carbon reserves in the Arctic, was the idea that it would be wise to create a "climate change rapid response team."

Walter's haunting image of a leaky cryospheric cap is a compelling one. What she describes is an Arctic permeated by carbon – a veritable sink of biogenic and geological sources of methane leaking into the atmosphere, and the only thing that is preventing wholesale release and runaway warming feedbacks is winter cold that slows the process and keeps the leaky cap of permafrost partially in place. That cap, however, is breaking down. Though Walter hasn't said it in so many words, what we are facing in the Arctic is a carbon release time bomb. Only by keeping the Arctic cold can we prevent that bomb from going off.

FROZEN SUBSEA METHANE

A related problem is the amount of methane trapped in the cold marine sediments of the polar oceans. In their book *Methane Hydrates in Quaternary Climate Change: The Clathrate Gun Hypothesis*, authors James Kennett, Kevin Cannariato, Ingrid Hendy and Richard Behl explain how methane comes to be stored in polar oceans and why warming that leads to the mobilization of this gas could destabilize the global climate.

Biologically generated methane is almost exclusively produced as an end product of metabolism by a large group of primitive anaerobic microbes called methanogens. Among the most ancient forms of life on Earth, these microbes only exist in the absence of oxygen, which to them is a deadly poison. Methane production, or methanogenesis, can therefore only take place in water or under other conditions defined by the absence of oxygen. Methanogenic microbes can produce huge volumes of methane under the right anoxic conditions, which are found typically in the sediments of freshwater lakes, marshes (where it is often called marsh gas), in rice paddies, flood plains and tundra peatlands and in marine sediments where organic carbon is available in sufficient concentrations.

Methane hydrates – or clathrates as they are known – are the ice-like solid form of methane produced through methanogenesis. This solid methane is produced when methane molecules are captured within a cage of water molecules at low temperature. Hydrates form from other gases as well, but the methane variety is the most abundant and widespread, especially along the continental margins and in shallow marine basins throughout the northern polar regions.

Kennett and his co-authors point out that methane can also be generated at greater depths in these circumstances by thermogenic processes. They report that methane production from deeply buried organic matter can occur at temperatures from -80°C to +150°C. Methane formed in this manner can then migrate vertically through the sediments to contribute to methane already trapped in clathrates or be released directly into the atmosphere via seawater.

Given the geographical scale of anoxic environments globally, it is not surprising that the volume of methane stored in hydrates is very large. Though they remain uncertain and speculative, global estimates of the amount of methane present in clathrate form suggest there is probably 3000 times more of it stored in this form than presently exists in our planet's atmosphere. Equivalent or even greater volumes of methane are thought to be trapped below the hydrate zone.

Clearly the volume of methane trapped in clathrates and as associated free gas is sufficiently large to play a role in global climate change. The problem of methane release from cold-water clathrates may be orders of magnitude more significant than the release of methane from thawing permafrost. Marine geologists and climate scientists have determined that the release of marine sedimentary hydrates may have been the cause of rapid climate warming in the Earth's past. It appears these catastrophic releases were caused by sea level rise and warming temperatures in waters of intermediate depth over upper continental shelves in the circumpolar North – exactly the conditions that have come into existence in our time.

THE POLITICAL CHALLENGES

THE NORTHWEST TERRITORIES: A RIVER RUNS THROUGH IT

If the Columbia is the Great River of the West, the Mackenzie is the Great River of the North. In 2009 the Rosenberg International Forum on Water Policy reported that in its estimation the Mackenzie basin was important not just to the eco-hydrological integrity of the Northwest Territories, but to the world.

The Northwest Territories is almost entirely contained within the watershed that contributes to the Mackenzie River system. Though this is the largest river in Canada – larger even than the St. Lawrence – few Canadians have ever seen it. That does not mean, however, that it is not important. The Mackenzie basin is thought to be one of the linchpins holding North America's water-ice-climate interface together. Scientists believe that if the stability of this important eco-hydrologic system were to become compromised, it could cause the Earth's climate to wobble further out of its current equilibrium, with implications for all the ecosystems on the continent whose stability is coupled to current climate variability – which include many ecosystems in southern Canada. Such concerns are clearly not being taken lightly in the Northwest Territories. Over the past decade the people there and their territorial and Aboriginal governments have found themselves confronted with hydro-climatic changes which they cannot easily adapt to and can no longer ignore.

THE PEOPLE TAKE ACTION

The people of the Northwest Territories do not want to wait to see if methane releases from thawing permafrost or melting clathrates would result in runaway global warming. Neither do they want the landscape they live on to have disappeared from beneath their feet before they started to act. They recognized that one of the only immediately effective ways to adapt quickly to the growing number of negative consequences and costly feedbacks associated with a rapid change in climate is to manage water effectively. In order to assure their way of life, major water policy reform was clearly necessary.

Such reform is not easy, but because there was less debate in the Northwest Territories about water than about many other resource-related issues, it was agreed that affirmation of a new water ethic could be a means of ultimately achieving greater adaptive capacity to climate change while generating a great many other benefits along the way.

There was, however, a great deal of anxiety about how such reforms should be achieved. The NWT encountered all the same classic obstacles that southern jurisdictions face when they confront the need for water policy reform. The official departments responsible for water governance did not have adequate resources to deliver on full water policy reform. Jurisdiction over water was badly fragmented and institutions were often highly territorial with respect to their own powers. The entire system was bogged down under earlier precedents and projects. Both local and outside interests were invested in those precedents and did not want to interrupt a familiar and comfortable status quo. These issues are challenges everywhere, of course. But on top of this the NWT had unresolved indigenous land claims; the pending devolution of political powers to the territory from the federal government; and larger matters related to Arctic and northern sovereignty. There were also pressing competing issues such as the need for more and better housing, better education and more jobs to keep young people in the region.

Reforming water policy in the midst of all this seemed impossible. Concerns were openly expressed that, even in the face of obviously dangerous climate change threats to the region's social and economic future, existing political structures and institutions could prove incapable of dealing with such complex issues. It was realized that if this was in fact the case, NWT residents would have no choice but to accept that it was simply impossible to adapt their water management practices to growing climate change effects. This meant they were doomed to ongoing diminishment and loss of culture and landscape as their future birthright.

The people of the Northwest Territories, however, had too close a relationship with the place where they live, to accept such a fate. They realized that if they could manage their water differently, the way of life of most people living in the NWT wouldn't have to vanish. Fortunately a leader emerged in the form of environment minister and deputy premier Michael Miltenberger. Miltenberger and other Aboriginal government leaders worked together to make change possible.

So what did the people of the Northwest Territories do that allowed them to break out of the mould and change the way they relate to where and how they live? Unlike so many others today, the people of this region were out on the land enough to have first-hand experience of climate change effects they could see with their own eyes. They did not allow themselves to be prevented from breaking out of the prison of inaction by well-funded and carefully orchestrated efforts to deny or minimize the threat posed by hydro-climatic change. They recognized that what rising temperatures were doing to the function of the larger global water cycle was the place to start in understanding and acting upon the need to adapt to changing climatic conditions.

The people collectively agreed that diminishment and loss of culture and landscape were simply unacceptable to them. They came together around a common vision, which was to ensure not only that the water quality and flow of the Mackenzie would remain unaltered, but that their current way of life and relationship to place could remain unaltered as well.

THEIR GOVERNMENTS FOLLOW THROUGH

The story might have ended right there, with fine words and good intentions, but the prospect of losing their way of life was completely unacceptable to the people of the Northwest Territories. Governments at all levels acted. Departments responsible for water governance found adequate resources to deliver on full water policy reform. Earlier precedents and projects were embraced with reforms. Attention was and continues to be paid to unresolved land claims, and pending devolution of political powers to the Northwest Territories is now well advanced.

But it is in water policy reform that the NWT truly succeeded. The Northwest Territories became the first provincial or territorial government in Canada to clearly and legally define adequate access to clean water as an inalienable human right. Next, the territorial government and its Aboriginal and federal partners set out to craft its landmark *Northern Voices, Northern Waters* water stewardship strategy.

It has been observed that this stewardship strategy is a precedent-setting document nationally and globally because of the way it was crafted and because of the degree of accountability and transparency it demanded of everyone involved in its implementation. What is unique about the strategy is that all Aboriginal and community governments in the Northwest Territories *expected* to be involved from the outset in the development of the strategy and throughout all implementation and evaluation phases. While it was expensive and time-consuming to develop the strategy in this way, even at times frustrating to all participants, the collaborative manner in which it was crafted is the key to the strategy's success. In the case of the Northwest Territories, co-development of the strategy means co-implementation.

It is also important to note that the *Northern Voices, Northern Waters* strategy document was guided by a holistic ecosystem approach based on watershed-scale management ideals. Within this framework, effective monitoring and research programs of the kind funded by the Canadian Foundation for Climate and Atmospheric Science were identified as critical to the successful implementation of the strategy. This element has already yielded important results. When, in 2011, federal Environment Minister Peter Kent announced the closure of 21 of 23 monitoring stations in the Mackenzie system, the territorial and Aboriginal governments of the NWT were able to point to their water strategy as a reason for reversing that decision. This in itself is evidence that the NWT approach is already being integrated into other government programs that promote or overlap into water stewardship. These include drinking water quality frameworks; municipal water management practices; land-use plans pursuant to land claims; hydroelectricity and transmission plans; and protected areas, fisheries and greenhouse gas strategies.

WHAT MIGHT CANADA AND THE REST OF THE WORLD LEARN FROM THE NWT EXAMPLE?

The Northwest Territories has demonstrated to the rest of the country and to the world that through hard work and persistence of vision a relatively small population with limited resources and only a tiny budget can effect meaningful change. Doing the right thing is not a question of austerity or affluence but of priority and focus. It is a matter of leadership at all levels working together. It is no longer possible anywhere to say it can't be done. The Northwest Territories just did it. But the NWT story does not end there. Their real success only begins with the *Northern Voices, Northern Waters* stewardship strategy.

It has been the dream of many Canadian water experts that short-sighted southern water policy would not only be stopped dead at the NWT border, but its very contact with the holistic principles to which the Territories aspired would become the foundation of water policy reform that would radiate back south to positively affect all of Canada. Miraculously, that appears to be exactly what is happening.

Somehow, the Northwest Territories summoned the courage, administrative capacity and scientific wherewithal and claimed the necessary moral ground to permit them to negotiate the kind of progressive transboundary agreement it signed in March of 2015 with upstream Alberta. It is this agreement that is paving the way for similar pacts with the Northwest Territories' other riparian neighbours.

There were a number of criteria and principles upon which the NWT strategy was founded that were deemed worthy of consideration in provincial and other jurisdictions in southern Canada. Though it was so obvious that at first it was missed, the most important thing said about the NWT water strategy is that it demonstrated that governments at a number of levels can do exactly what governments are supposed to do when faced with a serious societal threat: govern. Those agencies responsible for the development of the strategy took their fiduciary and legal responsibilities for the management of water and watersheds in the Northwest Territories seriously. They did not back away from the project because water policy reform on this scale was too difficult to agree on or too politically sensitive to address, as so often happens in other jurisdictions. They did not bow behind closed doors to individual sectoral interests or hold back on the extent of reform in response to special pleading. They did what so many governments never seem to manage: take charge, collaborate with affected constituencies and find and implement solutions that hopefully will work over the long term. The NWT did what was necessary, not what was easy. The territorial governments got the whole job done, not just the parts of it that would have been an easy short-term political sell.

THE MACKENZIE RIVER BASIN BILATERAL WATER MANAGEMENT AGREEMENT

In March of 2015, leaders in the Northwest Territories capitalized on the strength of *Northern Voices, Northern Waters* to establish new and enhanced transboundary relations with one of their upstream neighbours based on the principles of the strategy. With Merrell-Ann Phare as chief negotiator, the NWT team crafted a landmark agreement with Alberta relating to the future joint management of the Mackenzie basin. This agreement – the first in the country to institutionalize a co-operative adaptive response to changing eco-hydro-climatic circumstances over time – should be a source of lasting pride and inspiration to the two signatory jurisdictions and to the nation for decades to come.

The pact is remarkable in a number of ways. Its foundation rests on the same depth of Aboriginal input and accommodation concerning rights and the perpetuation of traditional ways of life as the water stewardship strategy whose values and structure it reflects. The agreement includes provisions related to the roles water plays in ecosystem function, and to local cultural connection with those roles. It is also similar to the Northwest Territories water strategy in that it was collaboratively crafted and its conditions are such that people who live in or care about the Mackenzie basin can hold governments accountable for honouring not just the letter of the agreement but also its principles.

Like all transboundary agreements, however, the Northwest Territories–Alberta pact will require considerable ongoing collaborative attention if it is to meet its objectives. This agreement and the other, complementary ones currently being negotiated with British Columbia, Saskatchewan and Yukon rely heavily on keeping a finger on the pulse of eco-hydro-climatic circumstances throughout the Mackenzie basin. This will require a long-term commitment by all of the jurisdictions in the Peace–Athabasca–Mackenzie system to careful, precise monitoring and effective, meaningful and timely sharing of data. The test of this agreement will be the extent to which resources are allocated to meet these conditions now and over time.

The national and international significance of the NWT *Northern Voices, Northern Waters* stewardship strategy derives from the patient and inclusive manner in which it was crafted and implemented, and from the ongoing engagement through which its effectiveness will continue to be thoughtfully monitored and evaluated. In that it respects traditional knowledge and ways of life in the context of basin-wide aquatic ecosystem health, the agreement represents a landmark in integrated watershed management. Its example demonstrates that in fact there are no real jurisdictional, legislative, constitutional or political obstacles to the creation of a sustainable future. All it takes is political will and competent, persistent leadership.

In 2015 the Northwest Territories made history. And because the NWT stuck by the principles implicit in their own strategy, everybody in Canada has won. By their example they have shown that the road is open for reconciliation in this country; for correcting the mistakes of the past; and for collaboration on a better, more secure, more just and ultimately far more sustainable future for all. What the NWT has done is exactly what the rest of Canada needs to do: they have created a practical water policy framework that will provide hope for a sustainable future for all who live in the Mackenzie basin.

Analysis of the Northwest Territories response to the need to adapt as soon as possible to climate change effects on water, however, leaves us with more questions than answers. First, shouldn't we be supporting the kind of scientific research that will allow us to understand and proactively get to work on the kinds of hydro-climate changes that are beginning to appear and that according to all climate models are likely to accelerate everywhere in Canada in the future?

Shouldn't we in the south be supporting and emulating the Northwest Territories in its efforts to achieve economic and environmental sustainability in the face of hydro-climatic change?

Shouldn't we be looking at the parameters established in the NWT water strategy to provide overarching guidance in water policy reform in southern Canada?

Shouldn't we be cultivating the same kind of leadership provincially and nationally that has been demonstrated in the Northwest

Territories, so that we can be sure that water quality and availability issues associated with climate change don't limit our social and economic future?

Isn't it time for a nationwide water strategy?

CHAPTER FOUR

CLIMATE AND THE CRYOSPHERE

The Snows of Yesteryear and the

Future of the Mountain West

If, dear reader, you've come this far into this book but are still not sure exactly where you stand on the climate issue, I fully understand your position. I have worked for more than a decade on communicating scientific outputs from six climate research networks, and after reading hundreds of scientific papers and reviewing more than 150 books on climate change, there are only three things I am relatively certain of at this moment. Notice, however, that I said I was only *relatively* certain, which suggests that, like the scientists I rely on for my information, I am always open to new perspectives, provided they emerge from credible sources that have had the benefit of being evaluated by others with equally credible reputations for scientific objectivity.

I presently subscribe to three views. First, carbon dioxide concentrations in the atmosphere are increasing rapidly, causing mean annual temperatures to rise. Second, we humans are the principal cause of these temperature increases. And third, the warming atmosphere is becoming more turbulent, resulting in widespread weather pattern disruption, more intense droughts, extreme floods, bigger fires and wilder storms all over the world.

I have not always been this confident on any of these issues. I too have been plagued by doubts. But over time I have had the benefit of examining and testing these doubts in association with some of the world's leading climate scientists. Working among these circles, I can say with confidence that no water or climate scientist I know wants to see what is happening to the planet's atmosphere continue. We have changed the composition of our atmosphere by way of our collective emissions and now we have to address the consequences of global change. This is not a minor proposition. Our future depends on how we respond now. The full weight of the entire scientific method is being brought to bear on determining the facts so they can be translated into appropriate action.

The scientific method is highly structured. Even when known deniers and cranks challenge any aspect of the growing body of climate science, researchers are bound by the method to test the claim against the entire body of established knowledge to see if it deserves consideration. This is how science spirals gradually forward over time. Science is never "settled." Every critical objection to the hypothesis that our global climate is warming and that the burning of fossil fuels is contributing to this warming has been and will continue to be subjected to analysis and fierce scientific debate. So far, however, what has consistently happened is that after running objections again and again through the entire climate science knowledge system, we keep arriving back at the same conclusions. Yes, of course there are other factors besides carbon dioxide that influence climate. Climate is complex. But when you put all those factors together their influence doesn't even begin to explain current warming.

We now know that the output of the sun has not been a significant influence on global temperatures for the last 50 years. Careful research has concluded that changes in the Earth's orbit will not appreciably affect atmospheric temperatures for probably another 30,000 years. Our fossil fuel burning is presently pumping out CO_2 50 to 100 times faster than volcanoes do. Yes, modern volcanoes emit only 1 to 2 per cent of human emissions. Yet despite the localized cooling effects of La Niña events and other ocean current influences, average global temperatures continue to rise. Yes, warming will eventually decrease CO_2 concentrations by way of weathering, but that cycle takes half a million years to kick in.

With each passing year and each research outcome we keep arriving at the same conclusion: it's us – we are the cause. Despite this, there are people who continue to insist that the atmosphere is too large for humans to influence. As mentioned earlier, though, the Earth's atmosphere is only a fraction of the volume of our oceans, and look what we have done to those. It has been estimated that we have dumped perhaps half a trillion tonnes of carbon dioxide into the atmosphere. For purposes of comparison, let's imagine pumping even half that amount – 250 billion tonnes –into our oceans to see if anything will happen. Of *course* something will happen. We may not be exactly sure *what*, but we know *something will happen.*

Nor can you say that the tiny percentage of carbon dioxide in the air makes its presence meaningless. We know that carbon monoxide is fatal at 35 parts per million, which is less than a tenth of the current concentration of CO_2 in our atmosphere. Cyanide gas is fatal at 135 parts per million after just 30 minutes. If there were the same amount of cyanide in our air as there is carbon dioxide, advanced life would cease to exist on land on this planet. If these comparisons don't inspire you, perhaps it would be appropriate to invite you to think of the impact a skunk can have on a campground or in your backyard. A small volume of atmospheric disturbance can go a long way.

The fact is that we have eliminated all other ways to explain the warming and are reduced to matters of basic atmospheric chemistry and physics. You can create your own physics – or dismiss physics outright or even deny science altogether as is happening in parts of this country and in the US – but you can't avoid the basic fact. We are changing the composition of our atmosphere, and the effects of that are becoming hard to ignore. If you wish to test the validity of this argument, visit your nearest glacier.

INVESTIGATING THREE COLD CASES

Under the aegis of the Canadian Foundation for Climate and Atmospheric Science, the Western Canadian Cryospheric Network, or WC^2N as it was known, orchestrated research at six Canadian universities and an American one into what is happening to our cryosphere – that part of the world characterized by ice and frozen ground. The objective of the undertaking was to clarify the fate of the glaciers in Western Canada. These scientists' work revealed dynamics that will enable further research to build on wc²n's findings so as to more accurately characterize exactly how much water might reside in the glacier ice of the mountain West.

PEYTO GLACIER

The area of Peyto Glacier in Banff National Park was about 14.2 square kilometres in 1949. By 2010 that extent had diminished to about 11.6 square kilometres, an areal loss of 19 per cent over the intervening 60 years. This is different from linear loss but directly linked to it. The volumetric loss of the Peyto over the same period is estimated at between 10 and 20 million cubic metres of water a year. Today the rates of melt are faster, the glacier itself smaller. Obviously, if these trends continue it will not be long before the volume of the Peyto will be so small that such melt rates will not be sustainable, a point that may have been reached in 2015.

It appears now, however, that some of the areal loss attributed to recession of the Peyto Glacier was in fact not loss at all. Though the lost ice disappeared from sight, some of it did not melt. Instead it became entrained in moraines, where it continues to sustain high annual discharge. The exact area of the Peyto entrained in ice-cored moraines is not known. It is exactly this kind of information that is missing from the outcomes generated by the Western Canadian Cryospheric Network. The same information is also missing from the evaluations of the current state of glaciers around the world presently being undertaken by the World Glacier Monitoring Service.

What is happening to Peyto Glacier is not unique. Periglacial areas are increasing exponentially while exposed glacial surfaces are being reduced linearly all around the world. The downwasting of the ice on the Peyto, however, is quite substantial. On average, the ice has been shrinking by some 4 metres a year, which explains why glaciologist Mike Demuth was able to toss a pebble onto the glacier surface from Peyto Hut in 1966 but would need a bionic arm to throw that far now. According to researcher Chris Hopkinson, both downwasting and periglacial entrainment are occurring simultaneously. This presents a number of problems. For example, when Geomatics Canada recreates maps, they may get the area of a given glacier right but they often use the same topographic data used in earlier maps to establish elevation. In most cases, however, the surface elevation of a given glacier will have changed because of downwasting – in some cases quite dramatically. If you are a researcher or maybe work for the Geological Survey, you might know that. But if you are a non-specialist just reading a topographical map, you might not.

Because of the long record of research into its dynamics, Peyto Glacier is used as a surrogate for other, similar glacier headwater situations. The dynamics of the Peyto, Hopkinson has noted, are different from those of surrounding glaciers. The nearby Bow Glacier, for example, at one time flowed over two big subsurface rock steps. It has recently experienced two periods of accelerated recession, one in the 1920s and '30s and another in the 1970s. Though the Peyto doesn't have the same underlying geology, it is anticipated that it too will experience episodes of step-like recession in the next 20 years. As the unglaciated area surrounding it becomes larger, the heat trapped by the greater area of bare rock and talus will overwhelm the glacier. This, researchers note, could happen quickly as change suddenly goes from linear to exponential in terms of the amount of area that becomes bare of ice and late-lying snow.

Chris Hopkinson has observed that while the exposed ice of the Peyto Glacier was downwasting at up to 4 metres a year, moraines were downwasting at a rate of 1.5 metres per year and then collapsing out onto the bare ice at a rate of as much as 6 metres a year, making it difficult to measure the full extent of glacial ice remaining beneath the rock and scree that covers its margins. Refining his earlier estimates, Hopkinson observed that mean wastage of the Peyto between 1949 and 2010 was likely to be in the order of 14 million cubic metres of water a year. Hopkinson's key point, however, was that moraine wastage and associated melt are now increasing exponentially as was feared. His main scientific concern was that none of these dynamics were presently being captured in the mass-balance record for the Peyto or for that matter any other glacier in the Rocky Mountains. The public policy implications of this are significant. It means we still have some way to go before we can accurately predict how much water will be released from the remaining glaciers in the western cordillera under a variety of climate change scenarios.

Hopkinson thinks he has a solution, however. Further thermal

sensing aimed at located ice buried under moraines would be conducted, along with LiDAR measurements of the entire glacier and
surroundings. LiDAR is short for "light detection and radar," and
its operating principles are similar to those of its much earlier sister technology, RaDAR, for "radio detection and ranging." With
LiDAR, however, the radio microwaves typical of radar are replaced
by near-infrared pulses emitted by a laser. This laser energy interacts with surface features and, upon scattering, returns to a detec-
tor. The calculation of surface area is a simple "time of flight" measurement, with the position of the survey aircraft accurately known
using the Global Positioning System. As a result of these observations it should be easier to include the dynamics of ice-cored moraines more accurately into future mass-balance calculations.

THE COLUMBIA ICEFIELD

Between 1975 and 1998, glacier cover as measured by area decreased
by approximately 36 per cent in the South Saskatchewan basin and
22 per cent in the North Saskatchewan basin. The loss of the glaciers at the headwaters of these two rivers is no small matter to the
three million people who live in the two basins and rely on that water. In the South Saskatchewan, where much of Alberta's irrigation
agriculture is located, water is already scarce and in some instances
already fully allocated. The North Saskatchewan basin is less water-stressed, but nevertheless is affected by annual average ice volume wastage equivalent to the amount of water used by approximately one and a half million people. Obviously, when there is no
ice left to become water, the growing needs of water-reliant populations, industries and agricultural sectors will exert ever-increas-
ing pressure on water availability.

The biggest challenge to properly assessing how much water is
stored in the glaciers in the western and northern cordillera regions
is the difficulty in accurately determining the exact volume of existing glacial masses. It is relatively easy to measure surface area, but
as noted earlier it is far harder to know how much ice is entrained in
moraines and rock falls. It is also very difficult to measure volume,
because it is hard to know the nature and character of the landforms
beneath the ice that define the volume of ice that can exist in any

given place. Because there is no simple and inexpensive way to do this, many of the early projections regarding the state and fate of Canada's remaining glaciers are only rough estimates based on the best information available as to the general topography of Canada's mountain regions.

In association with Parks Canada, the Geological Survey has funded preliminary research to precisely determine the volume of ice contained in the largest glacier system in the Canadian Rockies – the Columbia Icefield. Once popularly thought to cover 325 square kilometres, the icefield has recently been estimated at 223 square kilometres and shrinking. Now, however, the volume of the glacier system will no longer be merely approximated by estimating it as a function of area only. Instead the volume of ice will be calculated directly and exactly, through the use of ground penetrating radar surveys conducted on the surface of the icefield and by remote sensing using LiDAR. The aircraft carrying the LiDAR system will fly over the area, capturing information in swaths. Careful analysis of the data will yield an illuminated footprint of the landscape over which the detection of features and their exact relative distance from one another has been put into relief. The combination of LiDAR and ground-penetrating radar should yield the first truly accurate measurement of the form and thickness of the Columbia Icefield. By multiplying the amount of ice by its water equivalent it will be possible to estimate with a fair degree of predictive accuracy how long individual glaciers and even the Columbia Icefield itself may last under a number of projected climate change scenarios.

Given the central position of this icefield at the headwaters of three of the country's most important river systems, the Columbia Icefield Research Initiative, had it continued to be adequately funded, would have generated research outcomes of great value in estimating how much water will be available to future residents of the West. The Initiative could have been the most important scientific project undertaken in a Canadian national park since the parks system was created 130 years ago. A great deal of ice in the West is becoming water. By refining our understanding of the glacier ice that forms the hydrologic apex of Western Canada, this research could very well have defined the state and fate of the entire region. Out of this research we might have learned the hydrological limits of the dry West. We may have been able to better understand what the Western landscape can support, for it is water and not just natural resources that ultimately constrains sustainability on the Great Plains. It is water that makes the mountain and prairie West habitable. Financial support for the Columbia Icefield Research Initiative, however, was largely eliminated by the Harper government in 2014. Fortunately this did not mark the complete end of research on the rate of decline of glaciers in Canada's mountain West. But it did mean that the status of ongoing government monitoring and research is at the time of this writing uncertain and the work done at research institutions and universities has become more important than ever.

THE DEGLACIATION OF THE WEST

Garry Clarke of the Earth, Ocean and Atmospheric Sciences department at the University of British Columbia is one of Canada's most renowned scientists and one of the world's most respected glaciologists. In April of 2015, Clarke and his colleagues Alexander Jarosch, Faron Anslow, Valentina Radić and Brian Menounos published a paper in one of the most prestigious scientific journals in the world, on projected deglaciation of Western Canada in the 21st century. The paper, which appeared in *Nature*, confirmed earlier research conducted by many of the same scientists under the aegis of WC^2N concerning the rate and extent of deglaciation in the coastal and interior ranges of British Columbia and in the Rocky Mountains of Alberta. What was different about Clarke's and his co-authors' research was that it was founded on much improved and far more accurate models of glacial dynamics. Simulations of projected changes in the extent and volume of glacial ice in the region were also vetted against a broader ensemble of climate models.

The results of this research indicate that on average we can expect to lose, by the end of the current century, about 70 per cent of the glacial ice that existed in Canada's western mountains in 2005. The effects of this loss will be greatest in the interior ranges of BC and the Rockies and slightly less in the tidewater and coastal glaciers of central to northern BC. We can expect coastal glaciers to be diminished by 65 to 85 per cent, while losses in the interior ranges and the Rockies will be 90 per cent or more. Peak meltwater discharge into streams and rivers from shrinking glaciers will likely occur between 2020 and 2040. The projections of this research are that the main results of deglaciation will be changes in the hydrologic cycle in the region that will have consequent effects on water availability and the structure and health of existing aquatic habitat; on the timing and amount of hydropower generation possible; and on the diminishment of aesthetic elements that are at the heart of the region's recreation and tourism reputation. The loss of glacier ice will also affect water availability for cities, agriculture and natural ecosystem function.

The authors of the paper made specific reference to impacts that can be expected in the Columbia River basin, where the loss of glacial ice is expected to have effects on hydropower generation and flood control significant enough to have implications for the Columbia River Treaty reconsideration discussions that were initiated between Canada and the United States in 2014.

The takeaway from this research is that by the middle of the current century we can expect the Canadian West to be a different place. The people who live there should prepare to get along in that new reality. But there is a larger warning in this work that goes beyond the impacts associated with the loss of glacier ice. The same warming that is melting the region's glaciers will ultimately diminish snowpack and affect the timing and duration of snowcover, which will in turn impinge on water security in ways that will cascade through the social, economic and political fabric of the entire West. Snow matters.

"PURE AS THE DRIVEN SNOW"

As Martyn Tranter and Gerald Jones make clear in their chapter in the anthology *Snow Ecology: An Interdisciplinary Examination of Snow-Covered Ecosystems*, there is good reason to study snow, beginning with its chemistry. The chemistry of snow meltwaters can have a significant, if not definitive, influence on the nature and quality of surface waters. It has been found, for example, that snow generated in air masses originating in industrial and urban areas scavenges pollutants such as sulphur dioxide and nitrous oxides from the atmosphere. This produces acid snow. It has also been found that airborne pollution from operations like oil sands upgraders can result in highly toxic contaminants falling on and becoming part of regional snowcover. These pollutants can be flushed out of the snowpack early in the melt season, creating periods of relatively acidic conditions in soils and streams. As snowmelt continues, these contaminants circulate in higher concentrations throughout aquatic ecosystems, sometimes causing fish kills during the spring freshet.

Although pervasive, human activities are not the only influences on snow chemistry, however. Important processes also occur within the snowpack naturally. Researchers have demonstrated that snow is not a passive reservoir of chemicals. As snow ages and metamorphoses, its composition is altered as a result of chemical reactions within it. "Pure as the driven snow" may be a familiar cliché, but the sarcasm it's usually voiced with nowadays isn't too far off the mark.

When snow falls it brings down with it all the substances that become trapped within individual flakes during the descent, which means just about every kind of particulate that happens to be in the atmosphere at the time. Snow formation and snowfall absorb chemicals from the atmosphere in three ways: through imprisonment of substances in the initial stages of ice crystal formation; through capture of gases, aerosols and floating particulate matter that snowflakes pick up while falling through clouds; and through the scavenging of airborne contaminants as the flakes fall through the air below the cloud layer. Water droplets in clouds contain a broad range of solutes that come into suspension as a result of particulate capture and through the diffusion of atmospheric gases into the resulting liquid solution. These solutes normally include sea-salt aerosols, sulphur dioxide, calcium, potassium and magnesium.

The range of solids that come down in any given snowfall can also be astounding. If you have ever put freshly fallen snow in your mouth and tasted dust, you will some idea of just how much freight returns to earth in the nearly silent wafting of snowflakes. Tranter and Jones report that any given snowfall will contain elements that once formed the Earth's crust, such as the calcium and magnesium mentioned earlier, as well as terrigenous dust. There may also be weak organic acids, trace metals and whatever else was in the air at the time, including particulate matter from the incomplete combustion of fossil fuels and the out-gassing of volcanoes.

Tranter and Jones also recount that during the winter the chemistry of cold snow is influenced as well by the biological debris that falls on the snow from above or that comes into existence within the snowpack as a result of biological activity there. In our northern forests Canadians are familiar with the leaf and needle litter that covers the surface of aging snowpacks, particularly in the spring. The excrement of mammals such as moose, elk and deer that have crossed the snow also finds its way deep into the snowpack. Windborne arthropods and winged insects also fall to the snow surface, creating additional chemical sources that become nutrients for organisms already living in the snow.

Such organisms can be surprisingly numerous. Bill Streever's remarkable book *Cold: Adventures in the World's Frozen Places* cites a Canadian study that revealed nineteen species of spiders, fifteen of mites, sixty-two of beetles, sixteen of springtails, thirty-two of ants and wasps and two species of centipedes, all living together in a complex ecology in and under the winter snow.

Neither does falling snow always look the same, as any Canadian winter driver could tell you. Streever cites a report of snowflakes four inches across falling in Berlin and a case from Montana in 1887 when flakes were fifteen inches across and eight inches thick. Snow isn't always white, either: in 1815, when Mount Tambora in Indonesia exploded, it launched so much particulate matter into the atmosphere that snow later falling in Hungary was brown, in Italy red and in the eastern US blue.

It is well known that snow is one of the lightest substances on Earth and one of the best natural insulators. The insulation provided by snow often allows soils beneath to remain unfrozen. It has been observed that microbial activity does not cease in most soils until the temperature of the frozen soil drops below -8°C. Above this temperature, microorganisms produce trace gases by way of their respiration. While colder temperatures reduce the overall rate of this microbiological activity, the volume of trace greenhouse gases trapped beneath snowcover can still be substantial relative to annual natural emissions from the soil: up to 20 per cent of the carbon dioxide and up to 50 per cent of the nitrous oxide. Because snow is porous, these gases can concentrate in gas-rich air pockets within the snowpack. This is especially the case in alpine regions where the snowpack is particularly thick, providing excellent insulation for the soils underneath. Microbiological processes are sustained all year round in snow-covered alpine soils. In such situations, the snowpack also serves as a sink for carbon dioxide and methane. Changes in the depth and area of seasonal snowcover resulting from global warming have an effect on the trapping of these greenhouse gases, though just what the effect might be remains unclear.

It is not surprising, given how much chemical activity can go on within snow itself, that changes in snowpack and snowcover in the cold parts of Canada affect the very composition of the runoff that is produced in any particular region. Where snowpack and snowcover have diminished, the water available to streams and rivers will be chemically different from that produced by melt where the nature and extent of the snowcover have not changed. Cold, evidently, does not just matter to the amount and extent of seasonal snowcover; it also defines the very composition of the water generated in those regions where hydrological regimes are dominated by snow.

Chemical evolution in snowpacks begins as soon as snow falls, and continues right up until it melts. Certain chemicals are both created and lost as a result of physical changes that take place during the buildup and later dissolution of seasonal snowcover. Throughout the time snow remains on the ground, the fundamental chemical nature of the snowpack is transformed by meteorological conditions and in many cases by microbiological activity. Chemicals within the snowpack will migrate and may concentrate within the snowcover depending upon temperature and changes in the mass of the snowpack that occur as a result of sublimation or melting. The physical properties of individual chemicals play a role in determining the chemical loading of snowcover. As the snowpack melts, properties such as solubility and vapour pressure of chemicals scavenged from the atmosphere during snowfalls become important factors in determining the chemical composition of meltwater. It is important to note, however, that chemical transformations such as oxidization also occur within the snowpack, and the substances that fall with the snow are not always exactly what finally gets into the meltwater. As mentioned earlier, chemical transformations also take place as a result of microbial activity in the snowpack. Snow algae, for example, of which there are many species, assimilate nitrogen from within the snowpack. Other organisms in the snow also metabolize nutrients they find suspended in the snow column.

From all this complexity we begin to understand just how many parameters must be taken into account in order to predict how much water might be generated as streamflow in any given spring. We also get a sense of how many more parameters must be understood if a researcher wants to be able to predict what might happen in a snow-dominated system as the climate of that system changes over time. Modellers who wish to understand what changes in snowpack may mean in terms of the future security of supply cannot separate water quantity from quality. Changes in the volume or extent of snowcover will have clear effects on physical and biogeochemical characteristics of the entire environment influenced by seasonal snowcover. These effects will be significant far in advance of concerns over water availability for purposes such as dilution of treated wastewater or agricultural runoff.

The modelling of changes in the hydrological cycle also demands that processes related to such transport be parameterized. This is not a simple task. Though they may appear separate, the atmosphere and the surface of the biosphere form a single biogeochemical continuum in which interactions and exchanges between the two regulate and maintain the composition of the Earth's atmospheric envelope. To accurately model what happens over time to the snowdrift in your backyard – and what will happen to the contents of that snowdrift – one must understand the thousands of physical and chemical interactions that occur at the larger ocean-air–water–land–life interface and then be able to apply that knowledge on a local scale. Though techniques are improving, this continues to be a supreme challenge to the accuracy of existing predictive models. Such modelling will become even more difficult in the future in the face of rapid changes to many established parameters that are occurring as a result of the loss of relative hydro-climatic stability as a consequence of atmospheric warming.

Because we have changed the composition of the Earth's atmosphere, and energized the global hydrological cycle in doing so, we have inadvertently created a hydro-climatic time bomb in Canada that threatens to undermine our established sensibilities with respect to the reliable availability and quality of water we presently take for granted. As the next chapter will demonstrate, the hydro-climatic genie is out of the bottle and we don't yet know enough about the genie or the bottle to restore conditions to the relative stability we once enjoyed.

CHAPTER FIVE

THE GENIE IS OUT OF THE BOTTLE

Hydrological Stability Has Been Lost and There Are Consequences

As we have seen, the Canadian Foundation for Climate and Atmospheric Sciences funded three research networks that specifically examined matters related to water and climate security in Canada. These included the Western Canadian Cryospheric Network, or WC^2N, which examined climate effects over time on the depth and extent of glaciers in the western mountain ranges of Canada and in the Arctic. The second, the Drought Research Initiative, investigated the consequences of major drought on the Canadian prairies between 1999 and 2005, with the aim of characterizing future risks to our agricultural economy from climate change. The third network, called IP3, undertook research into the fundamental processes that function at the very heart of hydrology, with the aim of understanding how warming temperatures may be affecting the countless interactions within and between water in the myriad processes that together make life possible on this planet.

These research networks did not operate independently. A co-operative information and data-sharing arrangement was created so that what was learned through the micro-scale research by the IP3 network could inform the larger macro-scale hydrological analyses of both the glacier research network and the Drought Research Initiative. As a result of this collaboration it was discovered that Canada's hydrology is in fact on the move and that the established conventions by which our society typically characterizes these processes are no longer adequate where warming is changing both micro and macro relationships between land, life, water, weather and climate.

IP3 stands for Improved Processes and Parameterization for Prediction in Cold Regions. As the name implies, the research network devoted its efforts to examining what is happening at the most fundamental levels where water interacts in the environment to create the world as we know it. These interactions are known as hydrological parameters. They include, for example, the behaviour of atmospheric, surface and groundwater at various temperatures and in different types of terrain at various altitudes and latitudes during different seasons of the year in Canada. Parameters also include differences in the way water acts in various types of ecosystems, under different kinds of human influence and under a variety of solar and other conditions. In this context, researchers in the IP3 network examined subtle interactions such as those that determine how much snow gets stored on a given landscape; how it is distributed; its patterns of melt and rates of infiltration of the meltwater into frozen ground; how water that infiltrates into thick organic soils may act over time; and how all of these factors might influence streamflow and therefore affect water security.

IP3 also investigated parameters the average person maybe doesn't think much about, including the extent to which surfaces at various temperatures slow or accelerate evaporation; how seasonal differences in energy inputs from the sun affect the behaviour of both liquid and frozen water; and the biogeochemical effects of the length of time rivers and lakes are frozen in winter.

As well as being a lead investigator on his own projects, Dr. John Pomeroy was chair of IP3 and co-chair of its sister network the Drought Research Initiative. Because the hydrology of Canada is dominated by what happens to water in winter, Pomeroy explains, it is impossible to get an accurate picture of future water and security without a clear understanding of the extent of current water supply based on the weather systems that operate in cold regions at high altitudes and high latitudes. For this reason IP3 research was undertaken in nine river basins that represented the widest range of hydrological circumstances in the cold regions of the country. Work undertaken in these basins includes research into snowpack, snowcover and redistribution at Marmot Creek in Alberta's Kananaskis Country; analysis of how snowpack patterns affected the long-term presence of glacier ice on Peyto Glacier in Banff National Park; and research into subsurface water movement at the Lake O'Hara basin in Yoho National Park. The work also included ongoing studies of streamflow and other hydrological dynamics in the Wolf Creek basin in the Yukon and in the Havipak Creek, Trail Valley Creek, Baker Creek and Scotty Creek basins, all of which are located in the Northwest Territories. Additional hydro-meteorological research was also conducted under the auspices of IP3 at Polar Bear Pass in Nunavut and in the Reynolds Creek basin in Idaho.

While by definition IP3 focused principally on the Rockies and the western Arctic, the outcomes of its research were of great interest in other parts of the country, if only because there is no place in Canada where cold doesn't somehow matter in terms of weather and climate. There was another factor as well, which made IP3's research useful not just in Canada but in many other cold countries: IP3 aimed to contribute to better prediction of regional and local weather, climate and water resources in cold regions, including river basins in which streamflow is not presently gauged. Because of our relatively sparse population, there are many places in Canada where streamflow gauges either don't exist or once were present but have been decommissioned due to government budget cuts. Such research will be particularly valuable in other expansive and sometimes sparsely populated countries such as Russia, Australia and large parts of South America.

UNDERSTANDING THE WORLD BY MODELLING ITS BEHAVIOUR

The overarching goal of IP3, which ran from 2006 until its federal funding was discontinued in 2010, was to bridge the gap between what we know happens in the real world and what we can simulate through modelling. The difference between observation and simulation is held to be one of the critical issues in contemporary science, especially in the domains of water security and climate change prediction. Until we successfully resolve this, we will not be able to translate what we are observing through direct monitoring in the field into reliable forecasts and useful predictions of what might happen in critical hydrological scenarios as a consequence of a changing climate in the future. If we are unable to accurately simulate what we see happening right in front of us, using models that accurately represent what natural systems actually do over time, the predictions science makes about the future will not create a solid enough foundation upon which to base sound public policy on such matters as water security, infrastructure design and climate adaptation.

John Pomeroy maintains that the challenge is to choose a model that is consistent with the assumptions upon which you may wish to build. This means careful selection of the factors that influence the hydrological processes you want to simulate, so that resulting predictions mirror what is actually happening in nature. In cold regions this means modelling parameters must at a minimum include snow-melt dynamics, the influence of permafrost, how much snow is intercepted by forest canopies, and differential patterns of evaporation from bodies of cold water, frozen ground or glaciated drainage basins. These variables, Pomeroy insists, must be considered in addition to all the other fundamental hydro-meteorological parameters such as precipitation, runoff, evaporation, cloud formation, atmospheric advection and dozens of other energy and thermodynamic factors.

One does not have to be around scientists like John Pomeroy long to begin to understand just how complex nature really is. In learning which parameters are important it becomes clear that hydro-climatologists view the world differently. For them, nothing stays put and nothing can be contained in any kind of disciplinary box. Even the names of things change as the things themselves transform. Circumstance is nested within circumstance, reality within reality, with water performing different facilitating functions at each level.

The world we behold each day is not as solid as we think. It is in fact not so much a material thing as it is a complexly variable interplay of parameters great and small that only create the illusion of substance. Imagine a mirage. Reality is perfused with water and life made possible by all the ways in which water reacts with nearly every element in the physical world. Change a few parameters in that mirage and you change the world you see outside your window.

Even the best models available today can only reveal a glimpse of the extent to which water plays a role in every aspect of the construction and maintenance of our Earthly reality over time. Some parameters, however, have huge influence over the nature and function of any given hydro-climatic circumstance. The changing of a single main parameter – temperature, for example – changes all the others. It is as if temperature and its influence on water lie at the heart of our reality. If our global temperature changes, an entire new geometry is created around that change. Alter the temperature enough and a new world will create itself around that geometry. The biggest discovery of all is that this is exactly what is happening right in front of us. Rising temperatures have begun to change a vast array of visible and invisible parameters that define the very foundation of the world as we have come to know it. And it is happening here in Canada now.

Though the value of their work has yet to be fully realized by the public, Canadian researchers participating in the linked IP3 and WC^2N initiatives have collectively made astounding observations that have revolutionized our understanding of climate change in two crucial ways. The first observation is that new hydro-climatic parameters are emerging as a result of climate warming and that they are creating a new and different world. These parameters have to be understood if we are going to have any hope of determining the kinds of hydro-climatic conditions Canadians are going to have to adapt to in the coming decades. This means that entirely new mathematical descriptions of our cold regions have be developed if we are to be able to accurately predict the effects of climate change on our built environments or on the natural systems upon which those built environments depend for their stability and sustainability.

The second indisputable fact collectively observed by researchers funded under CFCAS is that hydrological systems everywhere in Canada are *already* changing. Climate change is no longer theoretical. Its effects are already being observed and clearly measured. While hydrological circumstances have changed only slightly in some areas, these changes are highly pronounced in the cold regions of Canada. The evidence is now incontrovertible.

By examining the parameters which define the dynamics of hydrological processes, researchers have made a profound discovery. We simply do not understand all of the hydrological variables that are operative in natural systems. What's more, many of the parameters we thought we had nailed down well enough to employ in models are no longer accurate, because natural system dynamics are being altered by rising temperatures. In other words, we are basing our models on what we know about a world that no longer exists.

Something has suddenly changed in our relationship to the world as we once knew it. What IP3 and WC^{2}N researchers are collectively telling us is that the principal concern related to climate warming is not temperature itself, but the effect of rising temperatures on water in its various forms. Unfortunately, this is not a message that is getting out. One of the reasons why is that social sciences have been at war with the physical sciences for nearly forty years over what constitutes the legitimacy of science. We have to change that.

THE WAR BETWEEN SOCIAL THEORY AND THE PHYSICAL SCIENCES OVER WHAT TRUTH IS

One could readily argue that the social sciences have undermined the physical sciences on the matter of climate change in at least three significant ways. The first major assault was mounted under the auspices of deconstructionist social theory, which holds that all human knowing is in essence relative and that there are few solid facts that cannot be subjected to questioning or made the object of differences in interpretation.

The principles of deconstruction were and continue to be employed to question the legitimacy of the outcomes of hard science in ways that lead, through a range of rhetorical tricks, to the questioning of the legitimacy of science itself. The deconstructionist outcome is that no one person's opinion – no matter how qualified the person may be – is ultimately worth any more than the opinion of any other. When applied to the climate change debate this means that opposing views continue to be held to have equal weight even if they are based solely on opinion that has no foundation whatsoever in observed fact. This has allowed our entire North American society to debate climate change as though it were an article of faith like religion rather than a fact that can actually be physically observed and measured. To the utter astonishment of the scientific community, the whole notion of even the thermometer has been deconstructed and found wanting. We are being told that whatever a thermometer might tell us bears only a contingent relation to truth. Its measurements can mean whatever you want them to mean. As a consequence, wishful thinking is permitted to carry the same weight as incontrovertible fact.

The second assault that social theory mounted on the Earth sciences was made possible through orchestration of techniques and tactics that fall generally in the domain of applied psychology. With deconstruction as a foundation, well-funded energy interests that wish to deny negative human influence on our planet's atmosphere have effectively employed the behavioural sciences to neutralize public discourse on climate change through sophisticated advertising and targeted public relations manipulation. The desired effect of this misuse of psychology was to create suspicion concerning the legitimacy of science as a means of accurately knowing the world, and to cast doubt on any consensus that may have appeared to exist within the physical sciences community regarding the causes and potential impacts of climate change.

Social theory's third attack on the physical sciences has been in the name of wealth. The vigorous misuse of psychology in the climate debate is made legitimate by the overwhelming contemporary influence of yet another social science in which all values are held to be arbitrary and relative: economics. In this field there are blistering debates on the comparative economic benefits of acting on climate change threats now as opposed to waiting to deal with the problem twenty or thirty years hence when the total wealth in the world will be greater and the hardship of action somehow reduced. While the field of economics continues to fail utterly to account accurately for the real, long-term environmental costs of today's prosperity, our society accepts its confident pronouncement that the wisest course we can take today is to discount the value of the lives of future generations relative to our own. Furthermore, economists tell us, we don't have to worry much about the global environment or about climate change, because market pressures and human ingenuity will carry the day. So vocal, ever present and powerful are the social sciences on such matters that the moment doubt is expressed about our planet's climate future, they suck all the air out of the room with their arguments. Starved of oxygen in the absence of any new fresh air, collective societal brain function is reduced, which serves perfectly to maintain the status quo.

The global thermometer, however, is as indifferent to the psychological tricks of advertising and unrelenting public relations as it is to economic theory. The temperature readings keep coming in, but no matter how many times we check them, the picture that is forming of our circumstances and our future doesn't change. If you are actually observing the world directly rather than indirectly through what others say they think is relatively true about it, the picture doesn't look good. Our world is warming up, and at the moment at least, it looks like it is going to warm up a lot more, with consequences for everything that is important, no matter where or how we live today.

Those who have manipulated social theory in their own interest and those who have based their perceptions and livelihoods on the picture of the world created through the misapplication of psychology and the misdirection of economics may well ask how anyone can say with any confidence that the world is warming. What I can now say in response to that question is that I appreciate the scientific method more than ever and trust its outcomes more confidently on matters related to climate change as a result of five full years of carefully examining the research outcomes funded by the Canadian Foundation for Climate and Atmospheric Science.

As discussed in chapter four, what I observed was that every time another scientist challenges any aspect of contemporary climate science – and this happens continually – researchers are compelled by the long-established conventions of the scientific method to take that challenge seriously. They must re-examine all the knowledge about climate that has been collected and validated to date so as to test the legitimacy of any given counterclaim.

Perhaps the best source of information regarding the implicate process by which science has come to consensus on matters related to global warming is Paul Edwards's book *A Vast Machine: Computer Models, Climate Data and the Politics of Global Warming.* Though it demands much of the reader, this book explains in great detail how the knowledge infrastructure related to climate science emerged from the field of weather prediction and how each challenge to what we know about our global climate system gets tested against that infrastructure. Edwards describes clearly how centuries of accumulated scientific knowledge are brought to bear on questions related to climate change and how with each challenge the entire existing knowledge infrastructure is "inverted" and recalibrated so as to ensure that no possible connection or implication is ignored in our efforts to truly understand the potential extent of warming and the impact it may have on our future.

The title of Edwards's book comes from a reference made nearly two centuries ago regarding the need for a global Earth-observing system that would allow us to accurately measure large- and small-scale changes in climate. In 1839, Edwards explains, a young British art and cultural critic named John Ruskin dreamed of "a vast machine ... systems of methodical and simultaneous observations ... omnipresent over the globe, so that [meteorology] may be able to know, at any given instant, the state of the atmosphere on every point on the Earth's surface." Today, Edwards points out, that "vast machine" is largely complete, constructed from components – satellites, computers, instantaneous telecommunications systems – that did not exist and could scarcely have been imagined in Ruskin's time.

Weather knowledge works, says Edwards, because its "vast machine" has been around a long time and has matured in its function through countless revisions of practices, countless generations of technological improvement and the endless efforts of forecasters to improve the consistency of data and the reliability of computations. Over time, the vast machine's many moving parts were lubricated with international standards, the introduction of new institutions such as the World Meteorological Organization, the development of computers and the emergence of digital media. Edwards points out that while much "friction" remains as to the reliability of data and the accuracy of computations, the predictive capacity of current weather forecasting has improved by many orders of magnitude since the days of 19th century telegraphy.

Our current climate knowledge infrastructure, Edwards notes, is built around and on top of weather information systems. But unlike the global weather forecasting system, which is now accepted as a reliable and trusted aid to transportation, public safety and increasingly to agriculture, the way we gather and interpret climate knowledge has not receded quietly into the background of our society's functioning. Quite the opposite, in fact. As Edwards points out, climate controversies continue to lead very publicly right down into the guts of the climate knowledge infrastructure where debates are revived again and again about the origins of particular data and how they are interpreted.

Edwards explains that there are a number of factors at play here. Beyond obvious partisan motives for stoking controversy, beyond disinformation and what Edwards believes is a very real "war on science," debates over climate science continue for a more fundamental reason. Weather and climate are very different. Weather is what is happening now or in the next five days. Standards exist for the collection of all data related to today's weather. Climate begins with patterns of weather that emerge over at least thirty years. In climate science, as Edwards notes, you are stuck with the data you already have, which includes numbers collected decades or even centuries ago. The people who gathered these numbers are no longer alive, so you can't ask them under what conditions they collected their data. Climate researchers are not only faced with "data friction," which can be defined as the struggle to assemble records that are scattered across the world. They are also confronted with "metadata friction," or the problem of recovering the context of the data's creation, which is to say restoring the memory of how those numbers came into existence, so as to determine their accuracy. Because climate knowledge today is subject to such high friction for both data and metadata, the contemporary climate knowledge infrastructure never disappears into the background. It is always in view because, as briefly mentioned earlier, it is constantly undergoing what Edwards calls knowledge infrastructure "inversion": continual self-interrogation, an endless examining and re-examining of its own past. For this reason the black box of climate history never stays closed. Scientists are continually reopening it and rummaging around for yet more information on how old numbers came into existence. This constant re-examination of how data was collected in the past results in new metadata, which in turn results in new data models and ever-changing interpretations of the past. This is why, Edwards explains, climate data is not crisp and perfect but oscillates around identifiable trends. As he puts it, this is why climate data "shimmer" rather than stand out in clear relief. This is also why differences of opinion proliferate even within a general scientific consensus on climate change.

Edwards's point is that if you understand why climate data shimmer, and always will, then you will understand why climate predictions too will always shimmer. Once that is obvious, it becomes possible to understand and accept the proliferation of slightly different views amid general consensus in the climate research community. Edwards argues that what is preventing us from accepting the fact that different views can in fact coexist within the general consensus is the outmoded Enlightenment ideal of knowledge as possessing or leading to perfect certainty. Another reason why we as a society don't understand how the climate knowledge system is being true to itself by not only tolerating but encouraging differences of opinion within the general consensus over climate matters is that our society is presently being influenced by a philosophical argument that has become fashionable in the past fifty years that promotes a widespread relativism that elevates virtually any skeptical view to the same status as expert judgment.

Edwards observes that in the 1960s and 1970s scholars in the emerging discipline of science and technology studies attacked what they perceived as "a technocratic elitism that elevated science above other ways of knowing and seemed to place scientists beyond the reach of moral values and democratic ideals." Edwards admits to being part of this movement, which sought to overthrow what he describes as "an internalist historiography of science that ignored larger contexts and questions of power." The object of this movement, Edwards explains, was to emphasize the fact that science was a human invention subject to what these theorists called the "social construction of knowledge."

Edwards points out that early on in this movement these notions appeared to serve a useful critical purpose. In his estimation, social constructivism correctly asserted that, whatever the explanatory power of scientific methods, scientific knowledge also depends on norms, values and aesthetic principles and on mechanisms of persuasion, challenge, agreement and evidentiary standards. Scientific knowledge therefore could not be reduced to mechanically applied methods or technical apparatus. Such methods are in essence inherently and deeply social. To understand how social values function in science, as in any other human endeavour, deconstructionist theorists argued that it was necessary to employ historical, sociological and ethnological approaches in addition to the scientific method.

It was at this point, Edwards explains, that things got a little scary. As practitioners of science and technology studies pushed these ideas further, Edwards reports, the notion that social processes are *necessary* to knowledge production sometimes blurred into the far more dubious claim that social agreement is *sufficient* for knowledge production. "Socially constructed" stopped meaning "built by people from natural materials" and started meaning something more like "negotiated collectively by social groups," full stop. Edwards explains that it was if people thought society could stop bothering with steel or glass and gypsum and just make skyscrapers directly from blueprints, mortgages and contracts. At an extreme, this view regards physical reality as unknowable and unimportant, and the history of science as purely contingent, or dependent upon something else for its relevance. Science became little more than ideology or groupthink within which any belief at all might come to count as "knowledge." As a result, Edwards began to observe, science and technology scholars all too often characterized all sides in a scientific argument as equally plausible, and began to see knowledge simply as the outcome of struggles for dominance among social groups. Any outcome, any knowledge – in their view – could always, one day, be overturned.

As Edwards notes, however, this strange social theory of knowledge soon began to weaken under the weight of its own internal contradiction. As Edwards puts it, it began to commit epistemological suicide. By depicting physical reality as inaccessible and insignificant even while taking social realities – people's views and their ways of influencing each other – as transparently and directly knowable, not to mention all-powerful, it began to rub the non-academic world the wrong way. Its relativism has come to be viewed as utterly useless for purposes of practical analysis, if not entirely incoherent. It appears that sometimes we can be too smart for our own good.

Unfortunately, however, a lot of damage was done before the deconstructionist fad began to wane and the terrible weaknesses in this kind of thinking were exposed as a dead end. Among the lingering, highly damaging results of the brief rise of social theory as it relates to science is that a corrosive suspicion of scientific knowledge still permeates our society.

It wasn't just science that was damaged by this kind of misguided ivory-tower abstraction. It is interesting to note that the author Wallace Stegner was dismissive of deconstructionists for many of the same reasons as scientists were. In a speech he gave near the end of his life, he lashed out at deconstruction and deconstructionist criticism as it affected contemporary literature. Biographer Philip Fradkin quoted Stegner in full on this matter in his *Wallace Stegner and the American West*:

> And one kind of critics, the deconstructionist magi of Yale and its colonies, have declared themselves superior to fiction, and to literature in general. It is a joke. Both it and its makers are themselves fictions, mosaics of culturally constructed myths, fragments of previous texts and stereotyped values, phatic verbiage, reverberations of old delusions, echoes of echoes. The delight of these critics is to show what shopworn stuff literature is made of and what shoddy tricks it employs. They destroy their own justification for being, and before long, like the snake that took hold of its own

tail and swallowed itself, they will vanish like all the other over-subtle and scholastic aberrations of history. Perhaps once we have had them, we will be immune, as we are after having had the measles.

In the closing pages of *A Vast Machine*, Paul Edwards notes that there have been signs recently of a return to sanity in the social sciences. Even some once ardent proponents of radical constructivism, Edwards observes, have reconsidered its wisdom. Harry Collins is a professor at Cardiff University in Wales, where he developed the Bath School approach to the sociology of science. Collins has traced the search for gravitational waves, and has shown how scientific data can be subject to interpretative flexibility and how social or "non-scientific" means can sometimes be used to close scientific controversies. Edwards reports that Collins recently proposed in an article in *Nature* that "the prospect of a society that entirely rejects the values of science and expertise is too awful to contemplate. What is needed is a third wave of science studies to counter the skepticism that threatens to swamp us all." Collins's "third wave" would recognize and respect the value of scientific evidence, the tacit knowledge gained from disciplinary experience, and the wisdom of expert communities. At the same time, this "third wave" would require of scientists that they "teach fallibility, not absolute truth" – recognizing the provisional character of all knowledge. This, it should be noted, is completely consistent with the ideals of the scientific method and with the notion of the proliferation of interpretations of knowledge within the general consensus regarding climate change put forward by Edwards in *A Vast Machine*.

The climate knowledge infrastructure, Edwards argues, "not only accepts the provisional character of knowledge but constructs its most basic practices around that principle." This is the meaning of infrastructure inversion with respect to the past and of model intercomparisons as they relate to the future. Edwards notes that the Intergovernmental Panel on Climate Change has already explicitly recognized the provisional nature of climate knowledge by accepting *controversy within consensus*, and by articulating the climate's past and its future as ranges and likelihoods, not firm black lines. He notes that since the mid-1990s the IPCC has reduced its use of quantitative expressions of uncertainty ("a 25 per cent chance" etc.) in favour of qualitative language ("likely," "very likely," "with high confidence" and so on), especially in its synthesis reports intended for a largely non-scientific audience. Such language, Edwards offers, communicates appropriate levels of trust rather than measurable "uncertainties" – which he considers a massively overused term that naturally invites a negative evaluation of knowledge quality.

Despite these improvements, however, Edwards is not optimistic about how long it might take for science to recover from the relativist attack of the social theorists. He does not see an end to the public relations onslaught that marks the undeclared "war on science" that is still very much in progress, particularly, though not exclusively, in North America. He questions the real value of social media in today's debate over global warming. While Edwards notes that blogs and citizen science do at least initially appear to increase the transparency of climate knowledge, they are not what they seem. On the surface, Edward observes, they look like another mode of infrastructure inversion in that they do extend "ownership" of the knowledge-production process, which can broaden consensus. But on closer examination, he notes, their effects are, so far, decidedly mixed. Some have contributed new insight, but at least as often such media promote confusion, suspicion and false information.

Edwards is clearly worried about the way public perception of climate change has been hijacked by self-interested public relations manoeuvres. He concedes there is a possibility that the global models contain some error of understanding, as yet undetected, that will account for the warming they all predict. Maybe, he says, their warming forecasts are just groupthink, with scientists influencing one another to parameterize and tune their models in similar but unrealistic ways. Perhaps, he admits, systematic errors in the data make global warming only apparent and not real. Maybe ClimateAudit.org, SurfaceStations.org or some successor project will one day prove that so many monitoring-station records were skewed by so much artificial heating that really there is no warming at all, or at least none outside the range of natural variability.

And couldn't it be possible too, he speculates, that the IPCC is just an elite club, just another interest group, representing a scientific establishment bent on defending its well-funded research empire? Could be, he concedes, but not likely. From his perspective such claims don't carry enough weight to be worthy of being taken seriously. There are just too many models, too many controls on the raw data, too much scrutiny of every possibility and too much integrity in the IPCC process for any of these conjectures to be even remotely likely. Edwards accepts that while knowledge once meant absolute certainty, science gave up on that standard long ago. Probabilities are all we have, he notes, and the probability that the skeptics' claims are true is vanishingly small. The facts of global warming are unequivocally supported by the climate knowledge infrastructure.

Does this mean we should pay no attention to alternative explanations, he asks, or stop checking the data? As a matter of science, he answers, no. If you are doing real science, you keep on testing every new possibility; in climate science, you keep on inverting the infrastructure. As a matter of policy, however, the answer is *yes*. In his mind the response is to bring the controversy within the consensus. You get the best knowledge you can, he advises, and then you move, or try to move, against the enormous momentum of the fossil-energy infrastructure on which the world depends. Not an easy task.

What I learned from Edwards's book is that engineering is a social science, subject to all the distortions and self-justifications that direct application to human situations causes in all applied sciences. If engineers are sociologists, then climate scientists are historians. Their work is never done. Their discipline compels every generation of climate scientists to continue to revisit the same data to correct previous misinterpretations or find some new way to deduce the story behind the numbers. Just as with human history, we will never have a single, unshakable narrative of the global climate's past. Instead we get *versions* of the atmosphere, a shimmering cloud of proliferating data images, convergent but never identical.

Edwards asks some important questions. Do we really need to know more than we know now about how much the Earth will warm? *Can* we know more? From about 275 ppm in the pre-industrial era,

Edwards notes, the carbon dioxide concentration reached 387 ppm in 2008 – its highest in 650,000 years. Moreover, the *rate* of CO_2 increase is also rising, from about 1.5 ppm per year between 1970 and 2000, to over 2.1 ppm per year since 2004. At the time of this writing, in April 2015, the carbon dioxide concentration in Earth's atmosphere stood at 401.52 parts per million and was continuing to rise steeply. It is now virtually certain that CO_2 concentrations will reach the doubling point of 550 ppm sometime in the middle of this century. By 2100, concentrations could be as high as triple or even quadruple pre-industrial levels, even under optimistic emissions scenarios. Then, on the last page of his book, Edwards wrote the most troubling thing I have ever read about climate impacts on the future I hope to leave for my children.

THE MOST TROUBLING THING YOU MAY EVER READ

Paul Edwards reports that www.climateprediction.net has run thousands of "perturbed physics" simulations, varying model parameters to find the full range of possible climate futures that models predict. From the results of these large ensembles, he notes, the leaders of that project have concluded that the actual climate sensitivity might be considerably higher than IPCC estimates – perhaps greater than 6°C. And that's just for starters, since the planet will almost certainly overshoot CO_2 doubling.

Even more important, Edwards points out that "these scientists speculate that *we will probably never get a more exact estimate of potential future climate change effects than we already have*, because all of today's analyses are based on the climate we have experienced in historical time." Edwards quotes Myles Allen and David Frame: "Once the world has warmed by 4°C, conditions will be so different than anything we can observe today (and still more different from the last ice age) that it is inherently hard to say when the warming will stop." If that is true, Edwards writes, the search for more precise knowledge has little hope of success. Even more troubling is the realization that implicit in this quest for precision is the notion that there is some "safe" level of greenhouse gases that would "stabilize" the climate. As Edwards notes, Allen and Frame's point is that we do not know this, we cannot at the moment find out whether it is true, and now there is ample evidence to suspect it is *not* true. In other words, there is no point at which stabilization of our climate will somehow miraculously occur. One might logically expect that this would be true also of projections of global population stabilization. Such projections rely on a great many assumptions that remain untested.

Edwards makes the case that our stakes in history can be high indeed. "From family, ethnicity and nation to holocaust, apartheid, slavery and war," he writes, "the facts of the past matter a very great deal. So it is with the history of climate, and the stakes have never been higher. Our climate knowledge is provisional and imperfect. Yet it is real, and it is strong, because it is supported by a global infrastructure ... Its large expert community long ago reached a stable consensus of the climate's sensitivity to greenhouse-gas emissions and on the reality of the global warming trend. That consensus has survived many rounds of intensive reviews from every imaginable quarter. We have few good reasons to doubt these facts and many reasons to trust their validity."

The climate's past and its future shimmer before us, Edwards explains, but neither one is a mirage. When the doubts of skeptics and climate deniers are sent down time after time through the inverted knowledge infrastructure of climate science, the system of knowledge testing regenerates the same conclusions time and again. Within a small range of disagreement there is consensus about what is happening to our global climate. The climate is warming and will continue to do so for some time. How much it will warm and how quickly we can't say and may never be able to say, until it actually happens. This is the best knowledge we are going to get, Paul Edwards concludes, so we had better get busy putting it to work.

TAKING CLIMATE CHANGE SERIOUSLY

Scientists have been telling us for some time that human activity is changing the composition of the atmosphere. Because we have never affected the atmosphere on such a scale before, even the best scientists can't tell exactly what is going to happen. But the fact that their predictions are coming true suggests we should be paying attention to the science. Interestingly enough, we find now that almost all of these changes have something to do with water.

What scientists have been saying for nearly ten years is that the hydro-climatic conditions that are emerging in response to warming are increasingly outside the established range of what Canadians have demonstrated an ability to adapt to over the last century. We are already beginning to experience increasingly frequent, deeper and more persistent droughts. Simultaneously – and somewhat counterintuitively – we are also beginning to experience the same intense rainfall and flooding that are becoming more common elsewhere in the world.

This increasing variability is likely to become even greater in the future, which will result in extensive and very costly ongoing damage, not just to public infrastructure such as roads, bridges, dams and water treatment plants but to our entire built and natural environment. The costs of adapting our infrastructure to these changes are presently incalculable.

The fear is that the cost of this continuing damage may in time be substantial enough to make it difficult to sustain prosperity as we know it today and still keep our cities, towns, national transportation systems and other crucial infrastructure in functional repair. Take Toronto, for example. It has endured three 1 in 100 year and six 1 in 50 year storms in the last 25 years. One of these storms caused nearly a billion dollars of damage in only two hours. The same thing is happening in other cities in Canada and around the world. It is here – with the immediate prospect of more intense floods and drought, rainfalls and snowstorms – that our climate woes could really begin.

NON-STATIONARITY

Whether we like it or not, climate warming is changing the way water moves through the hydrological cycle in many parts of Canada. To understand why climate change is such a threat to established stability it is important to understand the central role water plays in our planet's weather and climate system. The fundamental threat posed by climate change relates to what hydrologists call stationarity.

While few outside the specialized fields of hydrology and climatology presently understand the meaning of stationarity, it won't be long before the term enters public consciousness and the everyday vocabulary of people around the world. Why? Because the loss of hydrological stationarity means our world is changing, in some places faster than anyone expected and in ways that many will not desire.

Stationarity is the notion that seasonal weather and long-term climate conditions fluctuate predictably within established and predictable averages. This perceived stability permits us a relatively high degree of certainty when it comes to predicting and managing the effects of weather and climate on our cities and our agriculture. It suggests to us, for example, that winters will be only so cold and summers only so hot. Stationarity induces us to believe that melt from winter snowpacks will always contribute roughly the same amount of water to our rivers each year and that rivers will rise only so high in spring and fall only so low in autumn. Because we know the range of what to expect, stationarity implies we only have to build storm sewers to a certain size, because we know from history that rainstorms only last so long and only result in so much runoff. Because of stationarity we can estimate how much water will reliably be available for hydropower generation or mining or hydraulic fracturing or other water-intensive energy projects such as oil sands.

Stationarity reassures us that sea levels will change only very slowly; that hurricanes will be of predictable intensity and duration and follow established tracks; and that tornadoes will only form in the most extreme conditions of the weather we have come to know and expect. This signals to us that we only have to build our structures to withstand winds of a known velocity, that our roofs will stay on if they comply with established building codes and that the cost of construction will be limited to what is required by those codes.

Stationarity also suggests that lightning will strike only so frequently and that water's symbolic and diametric opposite, fire, will burn only so much of our forests each year and will do it at times and at temperatures we can predict. From this we can infer how close to surrounding forests it is safe to build houses and that insurance policies can be arranged to cover those instances when the threat of fire exceeds known and predictable norms. In short, a reliance on stationarity statistics keeps the costs of disasters within limits we can predict and afford.

What is happening now, however, is that increased temperatures are altering the patterns of movement of water through the global hydrological cycle. This means that the statistics from the past as to how surface, subsurface and atmospheric water will act under a variety of circumstances are no longer reliable, which means we can no longer be sure that the volumes of water that were available to us in the past will be where and when we want them to be in the future. What IP3 researchers discovered, moreover, is that these statistics haven't in fact been reliable for some time.

Hydrologists observe that, in actual fact, stationarity in global hydrological conditions has likely never really existed, in that the climate is always changing, and that what we have actually done is established our own idea of the range of natural climate variability we think exists. We then constructed our society and all the infrastructure that supports it – including the risk assessments associated with maintaining that infrastructure – to fit within that range. Over time, this mistake in itself was bound to make us vulnerable, but now accelerated climate change is compounding that vulnerability and multiplying our risks.

The claim that we ought not to be concerned about this, because Earth's climate clearly has warmed to a similar or even greater extent in the past, is irrelevant in this context. The problem is that while our climate has changed in the past, it has not done so as quickly as it is changing now – at least not while there were seven billion people relying on out of date or poorly maintained infrastructure and living in stationary cities that were designed for climatic norms that no longer exist. Nor have humans experienced such changes at a time when the civilization they lived in was at the very limits of agricultural productivity and water supply globally.

By warming our climate we have torn open the fixed envelope of certainty within which we anticipated natural phenomena to fluctuate. In a very real sense the genie is out of the bottle, causing droughts, storms, floods and widespread pandemonium. As a society we have wasted four decades listening to arguments that these things wouldn't happen and that if they did we could easily adapt to them. And now we find ourselves almost completely unprepared for what is in fact happening.

The problem we face is that we do not as yet have an adequate replacement for stationarity statistics. In a rapidly changing climate such statistics become a moving target. Until we find a new way of substantiating appropriate action in the absence of stationarity, risks will become increasingly difficult to predict or to price. These risks, we discover, start small but quickly grow in terms of both potential impact and cost. Let us consider an example. How many culverts do we have in Canada? No one knows for sure, but they number in the millions. Under even the most moderate climate change scenario, how many might we have to replace? How about bridges? Roads? What will all this cost – and who will pay?

It is surprising, given these circumstances, that throughout Canada we are still thinking linearly about climate change. We still think the effects of it will be local, minor and cumulative, when in fact it will not be long before climate change will be affecting everyone everywhere simultaneously, compounding every regional economic, social and political disparity all at once. What we have collectively done over time is that we have quite accidentally – unwittingly, in fact – created a hydro-climatic time bomb. It is a bomb we have to defuse.

While the policy implications of this discovery suggest dark clouds forming in the direction of our future, this does not have to end up being an unhappy story. The implications of the ip3 realization that we have entered a period of non-stationarity do not by any means mark the end of the world. They only mean the world is changing.

If we are not able to contain the rate of change, the loss of hydrological stationarity will over time have serious consequences in many parts of Canada. The premise of this book is that we should be dealing seriously with emerging problems related to this loss of stability now, while time and prosperity are on our side, so that as our population grows and our climate changes, our social and economic future will not be limited by water quality and availability problems we could have and should have addressed in better times. What this discovery suggests is that it is time Canadians changed their attitude toward water.

The loss of relative hydrological stationarity means that more than ever we need baseline information, not just about the state of our water supplies themselves but of the basic eco-hydro-meteorological conditions that comprise the hydrological cycle as it functions in Canada. Over time there has been a serious decline in surface and groundwater monitoring as well as water research. While Canada used to be a world leader in these areas, cuts to government funding have resulted not just in a loss of leadership in these important domains but a great reduction in our own capacity to acquire and share water data and other information necessary to support climate change adaptation, especially in relation to our groundwater sources, of which less than 20 per cent are currently mapped. Commenting on this issue a decade ago, the Standing Senate Committee on Energy, the Environment and Natural Resources stated: "Clearly we cannot manage and protect that which we do not properly understand. When it comes to water, there are still too many questions to which we do not yet have satisfactory answers ... This information gap is more than regrettable; it is unacceptable. This stems in large part from the Government of Canada's retreat from water management issues and from funding relevant research."

More recently, the fall 2010 report of the federal Commissioner of the Environment and Sustainable Development echoed the Senate committee's stance: "... [Environment Canada] is not monitoring water quality on the majority of federal lands ... and does not know whether other federal departments may be monitoring water quality or quantity [there] ... The Fresh Water Quality Monitoring program does not validate the quality of the data it disseminates ... [thus the program] cannot assure that the monitoring data ... is fit for its intended uses ..."

Given the clear evidence that Canada's hydrological system is no longer stationary, the fact that we are now cutting rather than enhancing baseline monitoring capacity would appear to be nothing less than irresponsible. While there have been some improvements in various regions after recent flood disasters, the need for adequate monitoring still demands a great deal of attention.

We are not alone in facing these kinds of problems. Our neighbours to the south are facing challenges too, many of the same ones we will encounter as Canada's population continues to grow.

STOP POLLUTING THE WATER WE HAVE

In his 2011 book *The Ripple Effect*, Alex Prud'homme observes that the Clean Water Act in the United States was violated some 506,000 times between 2004 and 2009, but that less than 3 per cent of the violations resulted in any form of punishment. During the same period, the quality of drinking water in every state deteriorated as a result of violations of the Safe Drinking Water Act. One study revealed that the tap water in 45 states and the District of Columbia was contaminated by no fewer than 316 different pollutants, including the gasoline additive MTBE, the rocket fuel component perchlorate and industrial plasticizers called phthalates. Some 49 of these substances were found to be in excess of safe drinking water standards, meaning 53 million Americans had unsafe drinking water. Bottled water apparently did little better. One study demonstrated that bottled water purchased in nine states and the District of Columbia contained traces of 38 pollutants including fertilizers, bacteria, industrial chemicals, Tylenol and potential carcinogens. The point Prud'homme makes is that while industry associations may argue that such findings are exaggerated, we do not as yet know what cumulative effect water-borne compounds containing such diverse substances as sewage, plastics, pharmaceuticals, cocaine and Viagra will have on aquatic ecosystem health. All this is happening at a time when global demand for water will outstrip supply by 2030. Little, if anything, has changed in the years since these findings were published.

Prud'homme notes that the US Congress overrode President Nixon's veto to pass the Clean Water Act in 1972. This law limits pollution, sets water quality standards and penalizes violators. Right-wing political interests in the United States have been trying to reduce the CWA's power or eliminate it completely ever since it was passed. Prud'homme points out, however, that the biggest threat to water in the US is not huge oil spills such as occurred in 2010 in the Gulf of Mexico, but rather the myriad tiny leaks from cars, trucks, motorcycles, lawnmowers, boats, planes and other machines that rainwater routinely washes into sewers to end up in rivers, lakes and oceans. Prud'homme cites a 2003 National Research Council study that indicated that North Americans dump more than 300 million gallons of oil into their waters every decade, nearly double the highest estimate of BP's Deepwater Horizon spill in the Gulf of Mexico in 2010, which has been characterized as the worst environmental disaster in American history. The NRC report estimated that some 4 billion gallons, or about 15 billion litres, of oil leaks into the world's oceans every decade, more than 25 times the volume of the BP spill. As a litre of oil can easily contaminate as much as 5000 litres of water, this is not a small problem. As Prud'homme puts it, "the really big spill" is happening every day – all around us.

Neither cities nor bottled water companies, Prud'homme reports, test for pharmaceuticals. The Associated Press has estimated that the healthcare industry alone sends 250 million pounds of pharmaceuticals down the drain each year. The fear here is not just that hospital waste carries more antibiotics and germs than domestic waste. Researchers are also worried that the dumping of drugs into water systems could lead to genetic mutations that create antibiotic-resistant pathogens. This, Prud'homme reports, is not just dystopian paranoia. He cites research done in France at the University of Rouen Medical Centre which found that of 38 wastewater samples, 31 contained chemical compositions capable of mutating genes. Prud'homme also cites an Italian study that concluded that even trace elements of pharmaceuticals in drinking water – a few parts per trillion – "can significantly inhibit embryonic cell growth in vitro." Research in Germany confirmed the conclusions of the Italian work, as did later studies in the United States which also demonstrated that recycled water contains small amounts of pharmaceuticals and steroids – antibiotics, birth-control hormones, anticholesterol drugs, Valium and Viagra, among other things. It has been shown that some of these substances are resistant to even our most advanced three-stage water treatment processes, which has called into question the safety of using the recycled "grey water" that is increasingly mixed with natural supplies. One study of treated drinking water discovered high concentrations of a drug associated with the production of cocaine. Another in the United States found high concentrations of toxic substances associated with the illegal production of methamphetamines. In California, Prud'homme reports, poisonous chemical discharges from meth manufacturing have killed livestock, polluted streams and destroyed forests.

Prud'homme points out that while scientists have largely concluded that trace amounts of pharmaceuticals may not present a health hazard to human beings, little is known about the cocktail effect of many such drugs on aquatic ecosystem dynamics and on individual species, such as frogs that ingest Prozac, mosquitoes exposed to anti-seizure drugs or fish swimming amid what to them are physiology-altering concentrations of synthetic estrogen.

Studies in both North America and Europe have revealed that estrogen and other endocrine-altering substances exist in high enough concentrations to result in egg formation where the testes should be in male fish. In 2005 a United States Geological Survey researcher, Dr. Vicki Blazer, discovered that over 80 per cent of the male smallmouth bass she examined in the Potomac and other rivers in and around the District of Columbia were affected by endocrine-induced intersex characteristics. Since then, "feminized" male fish have appeared every year in the watercourses under study. A later, broader study that examined 16 fish species at 111 sites found that intersex affected a third of all male smallmouth bass and a fifth of all male largemouth bass. Research conducted by Leland Jackson at the University of Calgary showed similar effects on long-nosed dace. More and more evidence suggests that the principal cause of the intersex characteristic was synthetic estrogen from pharmaceuticals such as birth control pills, agricultural runoff loaded with pesticides and industrial drainage heavy with plastics.

Prud'homme cites a Canadian study undertaken in 2007 in the Experimental Lakes Area in Ontario where the renowned eco-toxicologist Dr. Karen Kidd added a small amount of ethinyl estradiol, one of the active ingredients in birth control pills, under controlled circumstances to a small lake. The introduction caused the feminization of most of the fathead minnows in the contaminated ecosystem. As a consequence, the minnows were unable to reproduce properly and the population was nearly wiped out.

The research conducted by Blazer in Maryland, Jackson in Alberta and Kidd in Ontario point to the same conclusions. If these substances are affecting the endocrine systems of fish, which are similar to that of humans, are we not now or could we not in the future face the same kinds of negative health effects as these fish?

Prud'homme points to additional work that suggests that manufactured chemicals released into our water and air may in fact be the cause of a surge in serious health issues such as breast cancer, leukemia, asthma, neurodevelopment disorders and physiological changes. He also notes well-documented findings that Western women, on average, are beginning puberty five years earlier than they did two centuries ago and are entering menopause later.

Prud'homme goes on to cite the work of Dr. Robert Hirsch, chief hydrologist of the US Geological Survey, who is of the opinion that a river is like a urinary tract. Just as physicians examine a patient's urine to determine what is happening physiologically in their body, analysis of river water – its chemistry, sediments and pollution – makes it possible to assess the health of a river's entire basin. That, Hirsch pointed out, makes hydrologists the nation's urologists. At a time of changing hydrological stationarity, this may prove to be very true – or at least it should be. Unfortunately, that is not turning out to be the case.

Environment Canada and the US Environmental Protection Agency share a common history. Both flourished in the 1970s, soon after they were formed. Since the late 1980s, however, both have suffered long, slow decline. Underfunded, overly politicized and widely dismissed as either toothless or irrelevant, each of the two environmental agencies suffered huge staffing cuts and became increasingly ineffective during right-wing administrations in the early twenty-first century. The EPA lost of great deal of its power and credibility during the two terms of George W. Bush, and Environment Canada lost similar ground while Stephen Harper formed successive minority governments followed by a majority in

2011. Both agencies today face the same limitations of jurisdiction and capacity in important areas related to water.

THE STATE OF OUR WATER INFRASTRUCTURE

The most pressing global water governance issue is infrastructure development, operation, maintenance and replacement. If we count the world's unserved – that is to say those who do not have reliable access to high-quality drinking water and sanitation – we are facing a multi-trillion-dollar water infrastructure deficit globally. In early 2015 the United Nations reported it would take investment of between $1.9- and $2.1-trillion a year for the next 20 years to supply clean water and sanitation to the billions of people presently without these services. This amount, it should be noted, is roughly equivalent to the subsidies provided to the oil and gas sector globally.

While we know that conservation is no longer an option, we have yet to satisfactorily define the water pricing and conservation measures that need to be put in place so as to free more water to be generally available for minimum human use and other purposes now and in the future.

The institutional mechanisms we have in place in Canada to pay for the orderly operation, maintenance and replacement of our crucial drinking water and sewage treatment infrastructure are inconsistent and won't hold up over the long term, especially as the hydrology of the country continues to change.

It would not be unreasonable to suggest that we have a three-tier system in Canada, each with a different set of rules. The first tier is the system that provides water to First Nations. That system is not working.

The second tier is the one that provides grants for infrastructure development in small communities and rural municipalities. Unfortunately there is no assurance that infrastructure grants will actually be directed toward crucial water infrastructure. Neither is there is any real incentive to manage such systems efficiently and sustainably. Success at this tier, therefore, is hit or miss. In this observer's estimation, operating this way is a recipe for growing infrastructure deficits. Because of the increasing presence of endocrine-altering substances and other new chemicals in our water, treatment standards are on the rise. Either you keep up or sooner or later you will face the political and legal-liability consequences of non-compliance.

Only at the third tier – where municipalities have priced water to ensure conservation gains and have instituted full-cost recovery and comprehensive asset management strategies – do we find the foundation for sustainability. But even in the domain of large municipalities there are problems.

Alex Prud'homme explores these big-city infrastructure issues at some depth in *The Ripple Effect*. Urban runoff, he argues, is even more daunting a problem than the pesticide and nutrient runoff facing rural communities. Because there is so much cement and pavement in cities, precipitation is not well captured. Moreover, the water-related infrastructure in many municipalities is old and leaky, and sewers designed for an earlier, more stable climate are easily overwhelmed by intense cloudbursts or prolonged rainfall. While cities do a better job than rural areas in separating human and industrial effluent, sewage treatment requires a great deal of energy, which adds to climate warming, which leads in a vicious circle to ever more intense storms. Most of North America's sewer network, Prud'homme points out, was built in the 19th or 20th century and is in poor condition. Contemporary urban water infrastructure here, Prud'homme notes, has been compared to a ruptured appendix – an overburdened system that is struggling to keep functioning and remain clean.

Not only are many urban water systems in desperate need of repair, many are also under assault from changing weather patterns. The sewer system for the City of New York, for example, is designed to accommodate a once-in-five-year storm – that is to say a rainstorm of an intensity that in the past was only expected to occur twice a decade. But weather patterns and storm tracks are changing. In 2007 alone, New York experienced three once-in-25-year storms – storms so intense they are expected to occur only once in a quarter of a century. The problem goes far beyond visible surface flooding, which obviously can cause tremendous harm in itself. Heavier than normal rainfalls also damage expensive infrastructure that you can't see because it is below ground. Such high-velocity flows can literally erode away whole sewer systems, as was demonstrated in Toronto in 2008 when a single heavy rain caused some $550-million worth of infrastructure damage in less than two hours. In such events, water treatment plants must be shut down or they fail. And when that happens, the combined outflows of domestic sewage, together with all the contaminants picked up by flood-. waters and carried through the sewer system, get discharged untreated into downstream watercourses, producing both pollution and public health risks that persist long after the flood recedes.

Municipal water managers have discovered that even with expensive improvements to sewage systems, there is no way to stop the effects of combined sewer outflows in extreme weather. The replacement of combined systems with "two pipe" systems that separate domestic sewage from stormwater outflows is astronomically expensive and often publicly unacceptable because such improvements, as again clearly demonstrated in Toronto, often require tearing up entire neighbourhoods. Prud'homme notes that the US Government Accountability Office has estimated it would cost $400-billion just to bring the nation's sewer systems back into fully effective function. No one knows what it would cost to upgrade those systems so they can respond to the new conditions created by changing weather patterns and more intense rainstorms. This means that for the foreseeable future at least, cities and the people who live in them are sitting ducks when it comes to extreme weather events which are expected to occur more frequently as our hydrological circumstances continue to change.

SAVING OURSELVES BY SAVING WATER

It could be said there is a silver lining in the discovery of non-stationarity in our hydrological cycles. We have taken water for granted for so long that we no longer think about how much we use or waste. We are among the world's worse water wasters. Under changing hydrological conditions we can no longer afford this luxury. Good can come out of this, however, if it stimulates water conservation.

One of the most startling revelations during the two years I spent travelling across the country examining and comparing water policy is how touchy Canadians are in defence of our culture of utter water waste. No one appears to be in any hurry to change, and today's political leaders by and large do not want to force anyone to do so. That is why reliance on engineering solutions to water issues is so popular. Even if such measures don't solve the real problem, they appear to make everyone happy. And that is what we want.

What has happened as a result is that we have accepted and encouraged wasteful water use as a social norm. We have, at enormous cost, overbuilt water infrastructure to support this wasteful norm. Now we find we cannot afford to maintain and replace all the overbuilt infrastructure that supports that waste, which increases the risk of public health disasters such as occurred at Walkerton, Ontario, in 2000. We have also discovered that we waste enormous amounts of energy treating and moving water. The cost of this energy is now rising and we are discovering we can't afford to spend up to 60 per cent of municipal energy budgets to keep doing this. In addition, we have realized that the energy we are wasting by wasting water is also accelerating climate change, which in turn is starting to pound the hell out of the infrastructure we can't afford to maintain and replace.

The situation must be seen for what it really is: an obvious vicious circle that is simultaneously bankrupting us while compounding climate change effects. This cycle will accelerate until we stop wasting both water and the energy it takes to treat the water and move it to the places where we waste it.

One of the biggest obstacles to change is that we have come to believe our own public relations spin. All across this country we have been telling ourselves for decades that our management of water is world class. Simply saying this, however, doesn't make it so. Strategic planning goals are just goals until you actually achieve them. One of the reasons we have been able to make exaggerated claims of our own success is that we largely compare ourselves only to ourselves.

At the time of this writing, average municipal water use in Canada was about 329 litres per person per day. Most of the countries in the European Union are at about 140 litres. Singapore is at about 125. The world champion right now, however, is Munich, at 100 litres per person per day. My question is this: Are we not deceiving ourselves just a little in claiming we are leaders when even our best average municipal water consumption is still three times that of the real leaders, who live in climates similar to ours in countries just as prosperous as ours and with a standard of living just as high?

The rest of the developed world has begun to act on the realization that municipal conservation is central to water and energy security, and that appropriate pricing is the most reliable means of achieving conservation of both. Recent statistics suggest we are now pulling ahead the US as the greatest water guzzlers in the world. In the middle of our continent is a vast semi-arid interior that most climate models say will likely get drier over time. Given our vulnerable circumstances, shouldn't we set our water conservation targets higher? Just the energy savings alone would be worth it.

The other thing we like to do in Canada is compare ourselves provincially to the rest of the country. Sure, any province can call itself a leader in water management in Canada, but what does that actually mean from water's perspective? Let's compare some of the provinces. British Columbia has existed as a jurisdiction for nearly 150 years, but it is only now introducing groundwater legislation. Saskatchewan has destroyed so many of its natural wetlands that it can no longer protect itself from flooding. Manitoba is now home each summer to algal blooms on Lake Winnipeg of up to 15,000 square kilometres, creating one of the largest freshwater dead zones in the world. Let's go to Ontario. After passing progressive legislation and spending billions to clean up industrial pollution, the province is now seeing the Great Lakes becoming contaminated again by agricultural runoff and urban stormwater pollution. Let's go to Quebec and talk water infrastructure. Montreal loses 50 per cent of its treated water to leaks. Its dilapidated water infrastructure puts it in the same category as cities in the developing world. Think of the energy waste. The Atlantic provinces too have these same problems to various degrees, plus one other: in Nova Scotia, New Brunswick and Prince Edward Island, sea level rise is an additional threat to water supply and infrastructure security.

Speaking here on behalf of water, one might ask if anyone in authority would wish to comment honestly and objectively on the state and fate of our country's aging pipelines? Let's look at ground-water across the country. In some places, contamination is so wide-spread that we are poisoning aquifers we share with the United States. But pointing fingers is not our purpose here. The goal is to understand that we are not doing as well as we think. Just saying we can do better is not enough. We have to *do* better, and there are sound economic and environmental reasons for that.

By conserving water, we can make more of it available for na-ture and for future generations of people. By conserving water we can reduce the size and cost of our water infrastructure and free up resources that will allow us to catch up on the maintenance of worn-out infrastructure. By conserving water we can save a for-tune in energy costs presently borne by the public to move water to where we waste it. By saving energy costs in wasting less water we reduce our greenhouse emissions, thereby slowing the very climate effects that are altering the hydrological cycle at the same time they are hammering our infrastructure. In this way a virtuous circle can be created in which we save water, save money, save energy and moderate climate change all at the same time. In so doing we also increase our capacity to adapt effectively to the changing hydrolog-ical conditions that are now emerging. This is simply too good an idea to ignore. In the end this could be a very happy story. Head-turning economic savings can accrue to governments and the pub-lic by way of water conservation. Breaking the combined water and energy waste cycle will save Canadians billions of dollars in direct infrastructure costs in the future while at the same time moderat-ing climate effects on the vulnerable infrastructure that exists today. Saving money by conserving water may be one of the most intelli-gent adaptations we can make to the climate-related loss of our na-tion's hydrological stationarity.

CHAPTER SIX

CONSIDERING THE STORM

Lessons Learned from the Western Canadian Floods of 2013 and Other Disasters

It had been three years since the Ottawa climate conference described in chapter two. The worst fears expressed at that gathering were realized when a spring flood inundated the town of Canmore, Alberta, in late June of 2013. During the night, a spectacularly swollen pulse of floodwater swept downstream from Canmore into the unprepared city of Calgary. Twenty neighbourhoods had to be evacuated and more than 100,000 people were forced from their homes.

And Calgary wasn't the only municipality in trouble. Twelve southern Alberta communities declared states of emergency, and eight of them besides Calgary were under evacuation orders. The flooding was particularly serious in small towns immediately to the south of the city. Waters at High River, for example, had risen so quickly that residents were trapped in their cars and homes and had to be rescued from their roofs. Two-thirds of the town was inundated and 5000 residents were forced from their homes.

With total damages exceeding $6-billion and a record $2-billion in insured losses, the Alberta flood of 2013 quickly became the most costly natural disaster in the history of Canada.

Late the following winter a conference was convened in Canmore to determine what lessons could be learned. But this was no ordinary townhall meeting. It was a high-level, expert analysis at which a great deal of significant scientific information would be exchanged. The conference was opened by the mayor of Canmore,

John Borrowman, who welcomed participants and thanked the organizers for holding this important conference in Canmore.

Introductory remarks by professor Howard Wheater underscored the importance of a newly established research initiative called the Changing Cold Regions Network (CCRN) in terms of understanding environmental change in the Rocky Mountains. Dr. Wheater summarized the goals, objectives and geographical reach of CCRN. He stressed the partnership aspect of the network, which would make it possible to establish observatories to provide data to help us better understand and predict regional and large-scale hydro-climatic variability and change. He noted also that the ultimate goal is to share new knowledge with a broader user community.

Professor John Pomeroy added his welcome and showed video of the flood event in the region. He then outlined the purpose of the workshop: to evaluate, analyze and synthesize the flood of June 2013 as a case study of extreme weather and hydrology. He explained the methodology of the workshop, which would be to compile a diagnosis of the floods and synthesize descriptions of the event and its implications, and thereby show how local floods are connected to broader atmospheric, hydro-meteorological and climatic trends. The workshop, Pomeroy explained, aimed for improved science and understanding of extreme-event processes that can be shared widely.

THE CONFERENCE GETS UNDERWAY

Professor Ronald Stewart led off with an atmospheric overview of the flood. His presentation, a prelude to later segments of the workshop, focused on climatology and the hydro-meteorological preconditions that led up to the flooding. Dr. Stewart provided a very detailed foundation for understanding how the 2013 storm formed, why it was so powerful and what was unusual about it. The event itself, he pointed out, was linked with a mid-level, closed low-pressure system to the west of the region and a surface low-pressure centre initially to the south. This configuration brought warm, moist, unstable air into the district, which led to dramatic, organized convection with an enormous amount of lightning in some areas, as well as some hail. Among the unique elements of the storm, he said, was the amount of lightning in the northern part of the region and the lack of lightning farther south, in Kananaskis Country. Lightning is an indication of strong convection – the lifting air current cools clouds and induces heavy precipitation; and the absence of such lightning in some of the main precipitation areas was notable for this region. Dr. Stewart pointed out that research is now being focused on comparing the devastating 2013 chain of events with other extreme weather to determine whether this storm was a harbinger of things to come.

Dr. Pomeroy followed with a summary of the hydro-meteorology of the storm. He noted that 15 researchers were in the affected area when the rain began to fall. He explained the geography of the area and showed a weather prediction model reanalysis of the precipitation that accompanied the event. Pomeroy noted that the storm was unusual in that precipitation did not increase with altitude in the mountains during the event. He pointed out that precipitation rates were not exceptionally high, but what was distinctive was the duration and areal extent of the rainfall. The effect of rain on snow was also important in generating additional snowmelt contributions to the rainfall runoff. At Dr. Pomeroy's study sites at Marmot Creek in Kananaskis, there had been a 41 per cent decrease in snow-covered area over the two weeks surrounding the flood period, which contributed at least 72 millimetres of snowmelt water to runoff during the flood. The limited capacity of frozen or wet soils to store water also contributed to the flash flooding. Together these factors resulted in the streams becoming torrents carrying heavy sediment loads. Pomeroy compared the 2013 event with episodes known to have happened in this area of the Rockies since 1879. Despite the heavy precipitation and concurrent snowmelt, the 2013 storm was only a 32-year return event at Banff and a 45-year return event at Calgary. For small mountain streams, however, there were no records that could put the flash flooding into historical context, as relevant hydrometric stations were destroyed in this flood and such stations did not yet exist for earlier floods.

Among Pomeroy's conclusions were that snowmelt over frozen ground contributed an additional 30 per cent to precipitation. We were lucky, he said. The Canadian Rockies flood was big but not extraordinary. It was not likely the flood of century as so many were calling it. Given the return periods of river flows, it was not even the flood of a lifetime.

Flood forecaster Colleen Walford then outlined Alberta Environment's flood forecasting mandate and methodology. She demonstrated how their team's forecast model works and sketched out the operational stages the team progressed through during the 2013 event. Walford concluded by outlining the reviews and projects that were set to be undertaken the following year aimed at improving the province's forecasting capacity for the present and into the future.

Bill Duncan of Saskatchewan's Water Security Agency introduced conference participants to the effects of Alberta's 2013 flood downstream in the neighbouring province. He showed the locations and outlined the storage capacities of dams on the South Saskatchewan River system, explaining how these works functioned to deal with the increased flows generated in Alberta. Duncan noted that despite careful operation of these dams, some flooding of farmlands did take place and unfortunately the Cree Nation at Cumberland House had to be evacuated. Major flooding was avoided at Saskatoon, however.

Continuing in a similar vein for BC, professor Sean Carey observed that the 2013 event also increased flows in the Elk River, which threatened the East Kootenay communities of Elkford, Sparwood, Hosmer and Fernie. With five active open pit coal mines in the area, mobilization of selenium and other contaminants was a concern. It was not possible to monitor the effects of such mobilization, however, because of safety issues during the storm and resulting flooding. Dr. Carey reported that rainfall in Fernie was 25 per cent greater than ever recorded. The good news was that defences constructed after a flood in 1995 proved largely effective.

Next up was professor Masaki Hayashi, who explained the potential roles of groundwater in mitigating or exacerbating the impacts of floods. He began by reminding us of the interactions between ground and surface water. He then cited research in the Himalayas that demonstrated the extent to which mountain aquifers can store water and buffer the effects of flooding. Dr. Hayashi introduced the hydrogeological research he and his students were conducting at Lake O'Hara in Yoho National Park. He also cited additional research at Marble Mountain, near Sundre, Alberta, where his team was able to employ an isotopic tool to calculate groundwater contribution to surface flows. This research demonstrated that up to 90 per cent of storm flows can originate as groundwater. Dr. Hayashi demonstrated how underground storage can both reduce and further contribute to streamflows during storms by introducing a delay in water flow. He then discussed the active exchange of Bow River surface and groundwater in the alluvial aquifer beneath Canmore. He noted once more that mountain aquifers were natural water detention systems. In conclusion, Dr. Hayashi asked a very interesting question: Can we utilize groundwater detention as a mechanism for flood risk mitigation?

The conference's lunch speaker was Don Cline, chief of the hydrology laboratory at the US National Weather Service, which is part of the National Oceanic and Atmospheric Administration, or NOAA. Dr. Cline's topic was extreme hydrology prediction. He began by noting that the National Weather Service has been providing flood warnings to the United States for 124 years, and that over that period there has been constant improvement in the quantity, frequency and lead time of forecasts. The recent increase in extreme weather events and the growing number of challenges in managing water, he said, have demanded that new tools be developed for prediction. As an example, Dr. Cline pointed out that his organization is moving toward ensemble hydrological models, like those used by climate scientists to predict future conditions, to make more accurate flood forecasts. The National Weather Service, he noted, has now expanded its prediction repertoire to include not only high flows and floods but also low flows and droughts. With its partner agencies, the service is now leveraging national investments in the direction of a nested Earth-system approach to understanding the global water cycle and its effects locally, nationally and globally, and is establishing a new 6000 square metre National Water Center in Alabama to further integrate and enhance US flood prediction. It did not go unnoticed at the conference that despite the heroic efforts of many of the scientists present, in many ways we here in Canada appear to be going in the opposite direction with respect to science-based flood and drought prediction.

After the lunch session, professor Julie Thériault examined the 2013 Alberta flood from the perspective of climatology, synoptic conditions and precipitation fields. Building on Dr. Stewart's earlier presentation, Dr. Thériault graphically explained the influence of the jet stream, synoptic forcing, accumulated precipitation, terrain, snowpack and snowline on the intensity and severity of the event. Noting that the source of water vapour for the storm was in the Gulf of Mexico, Thériault discussed the impact of increased sea surface temperature on moisture movement from the Gulf to Alberta. She also pointed out where efforts to improve current atmospheric models can be focused.

The next presenter was Logan Fang, who talked about the use of the Cold Regions Hydrological Model (CRHM) to simulate hydrological processes during Rocky Mountain floods. CRHM has been developed to model cold- and warm-season hydrological processes, including snow distribution, sublimation, melt runoff over frozen ground and unfrozen soils, evapotranspiration, subsurface runoff on hill slopes, groundwater movement and streamflows. No direct comparison between simulated and observed streamflow was possible, since, as mentioned earlier, gauging stations were destroyed in the June 2013 flood. However, basin records suggest that modelled peak streamflow lagged actual flows during the event. This suggests that the current model structure and parameterization are prone to previously undetected inadequacies in simulating peak streamflow timing during extremely wet conditions. The model was used to diagnose responses of hydrological processes in the 2013 flood for different environments such as alpine, treeline, montane forest and forest clearings in Marmot Creek in order to better understand flow pathways in such extreme wetness. To examine the model's sensitivity to antecedent conditions, "virtual" flood simulations were conducted using a week of flood meteorology (June 17–24, 2013) superimposed on the meteorology of the same period in other years (2005–2012) as well as in different months of 2013 (May–July). The results show sensitivity to snowpack, soil moisture and forest cover, with the highest runoff response to rainfall from locations in the basin where there are recently melted or actively melting snow-packs. Fang went on to note that, had the rainfall occurred in previous years at the same time of year, it would have generated a larger flood in most years. Had it occurred earlier in 2013, it would have produced a significantly larger rain-on-snow flood than occurred.

Bruce Davison of Environment Canada then offered a cautionary tale regarding the use of models to predict high-precipitation events. Davison noted that the Canadian Precipitation Analysis model underestimated precipitation at the basin scale, and he observed that model calibration is required to accurately parameterize precipitation, particularly in high-rainfall events. Flood simulations with the empirical WATFLOOD model showed great difficulty in simulating the flood peak on the Bow River without extensive calibration. However, the more physically based MESH model, which is a coupled hydrological land surface scheme that can be run directly coupled to Environment Canada's GEM numerical weather forecasting model, showed accurate simulation of Bow streamflow and peak flow during the flood. This was a very encouraging example of where better science and improved model physics lead to better prediction, and showed that Environment Canada's MESH coupled to GEM can provide the basis for a sophisticated national flood forecasting tool. Davison's presentation showed that the use of radar as well as rain gauges to provide reanalysis products from weather models can improve forecasting.

Professor Yanping Li then added to Dr. Stewart's and Dr. Thériault's earlier contributions with a presentation on the use of NOAA's weather research and forecasting model to simulate the 2013 Alberta flood. She began by explaining how the model worked and described the processes it could parameterize. She then revisited the synoptic conditions and water vapour sources at the time of the flood. This work, she explained, is being done to see whether model simulations can be used under global warming scenarios to determine whether these kinds of events might happen more frequently in the future.

Don Cline concluded the afternoon presentations with a detailed introduction to the US National Weather Service's Snow Data Assimilation System, or SNODAS. Dr. Cline's agency collects meteorological, satellite and snow survey data into this model, which estimates snowfall, snow depth, snow water equivalent, snowpack temperature, sublimation and snowmelt on a 1 kilometre grid across the US. He showed that the SNODAS satellite remote sensing program extends well into Canada. This began some time ago with US analysis of the Canadian portion of the Columbia River system. The most recent additions include all of Canada south of the 54th parallel. Dr. Cline explained that the agency did not just use satellite data and models but also relied heavily on field monitoring. It was very interesting to learn how much the current US data assimilation program relies on volunteers to collect such data. Also notable was the fact that the SNODAS website publishes virtually everything the agency knows about snow, much of it in three-dimensional imagery. In conclusion, Cline took conference participants back to June 2013 to demonstrate what the US knew about snow conditions in Alberta at the time of the flooding in the Canmore region. Utilizing remote sensing and the information available to them from field stations in both countries, they could tell snow-water equivalent, snow depth and snow melt in the Canmore–Banff area. From the SNODAS output they could see the potential for a substantial rain-on-snow event that was likely to result in catastrophic flooding.

In the evening, representatives of the Town of Canmore described the challenges the municipality faced during the 2013 flood and outlined their immediate response and short-term mitigation strategies. Before the flood, municipal officials had a limited appreciation of the risk, the size of the coming event or the damage it could cause, as no similar events had occurred in recent memory and flood plain maps were restricted to overflows from the Bow River. Through adaptive management and great effort, much infrastructure and all lives were protected in Canmore. The Town is interested in a risk management approach to mitigation, improved flood forecasting and debris flow prediction, and accordingly promised further consultation among academia, the provincial government, the consulting industry and the local community on possible solutions to the continued flood threat in Canmore.

CONFERENCE DAY 2

Opening the morning session, Dr. Danny Marks introduced partic-ipants to the Reynolds Creek experimental watershed in Idaho. Dr. Marks has analyzed the location of the rain–snow transition zone in the major storms that took place in the watershed between 1968 and 2006. In the 1960s, he noted, this area of Idaho had a snow-dom-inated hydrological regime. Today, rain predominates and this change has occurred in only 40 years. Marks observed that what causes melt in rain-on-snow events is not direct warming from the rain so much as heat flow to the snowpack due to energy released by condensation on the snowpack during rainfall. He went on to demonstrate how changes in snowpack energy can be modelled, noting that in large rain-on-snow events, snowmelt energy is in-creased by 50 to 100 times. Dr. Marks showed how changes in phase and energy balance affected snowmelt in a storm that occurred in Idaho over the Christmas holidays in 2005. He concluded by not-ing that if the 2013 Canadian Rockies rain-on-snow event had oc-curred when there was snow in the valley bottoms instead of just on the peaks, the flooding could have been orders of magnitude worse. For those who are concerned with future flooding in the Canadian Rockies this observation in itself is provocative. Should we expect more rain-on-snow events? With further warming should we also expect them to start happening in winter as well, as is occurring now only a short distance to the south?

Paul Whitfield expanded on Dr. Marks's research themes by de-scribing changes in autumnal streamflow in the broader Rocky Mountain region of North America, looking at a transect from Mexico to the Arctic Ocean. In a warming climate, he said, we should expect earlier snowmelt, lower summer flows, and changes in autumnal climate and hydrology leading to changes in precipi-tation patterns that could result in more numerous early autumn snowfalls turning into rain and rain-on-snow events. Dr. Whitfield noted that there is a trend toward increasing numbers of autumnal floods throughout the Rockies, something that is now clearly un-derstood in Colorado (see p. 132). Such events, he said, could be-come more common in this region of the Rockies as well.

Dr. Al Pietroniro then talked about the Water Survey of Canada and its role in basic measurement of water, with a special focus on flooding. He described Canada's national hydrometric system, the federal-provincial cost-sharing that supports it, and the challenges and opportunities of anticipating and responding to flood events in Western Canada. He explained how the Water Survey responded to the 2013 flood and how the course of the floodwaters was tracked through southern Alberta, Saskatchewan and Manitoba. He also showed the damage the flood did to his agency's monitoring station network and explained what is being done to restore the system.

Katrina Bennett spoke about the effects of extreme events on the warm-permafrost boreal forest region of the Alaskan sub-Arctic. She noted there is a trend toward warming in the interior of Alaska. While average annual streamflows are decreasing in response to this warming, winter and spring flows are increasing. Bennett pointed out that the 2013 river ice breakup was one of the latest in the season ever recorded in Alaska. March and April that year were warm in Alaska but May was very cold, delaying snowmelt. This late cold snap was followed by a record-breaking heat wave in June during which temperatures in some parts of Alaska approached 100°F, nearly 40°C. Her conclusions were that streamflow patterns changed in Alaska between 1951 and 2012. Systems dominated by snowmelt are declining in the spring, becoming dominated instead by rainfall. In the future, Bennett observed, Alaskans can expect more days with record-breaking maximum temperatures, with impacts that will cascade through the hydrological cycle.

Rick Janowicz built on Katrina Bennett's presentation with a description of the ice jams and freshet flooding that occurred during the same period of 2013 in neighbouring Yukon. Janowicz noted that the Yukon has a long history of flooding due to early settlement in flood plains along river transportation routes. He pointed out that 2013 was a cold winter and that snowfall in the Yukon that April was up to 250 per cent of normal. This was followed in May by a rapid snowmelt and runoff. Because of the cold winter, river ice was thick, and this combined with the compressed runoff period to cause serious ice jams. Janowicz recalled that flooding the year before had been so extensive it had cut off highway access to the Territorial capital, Whitehorse, which meant that food and other supplies had had to be airlifted into the city. Janowicz posed the inevitable question: Is this continual flooding connected to climate warming? He answered by noting that summer temperatures in the Yukon have risen on average by 2° to 6°C and that winter temperatures have risen on average from 4° to 6°C in the past century. Greater precipitation, higher temperatures, increased flows and compressed runoff periods all contribute to greater flooding. In concluding, Janowicz drew a knowing chuckle from his conference audience by pointing out that because of breakthroughs in computer technology, concomitant with a wider range of flood timings over the summer, it is now possible and occasionally necessary to forecast floods remotely even while travelling in foreign countries.

Dr. Roy Rasmussen concluded the morning session with a presentation about a major blizzard that took place in Colorado in March of 2003. Dr. Rasmussen and his colleagues modelled this storm in the context of what such a blizzard might be like if it were to occur in some future climate situation. The scenario they used was one in which mean temperatures were on average 2°C warmer with 15 per cent more moisture in the atmosphere than at present. The model showed that under these changed climatic conditions the current 10-year return period for blizzards of the intensity of the one that struck Colorado in 2003 would be cut in half. Rasmussen left the implications of such a change to the imaginations of the conference participants.

Model results showing heavier snowfall at high elevations were countered by ones showing more rainfall at low elevations and therefore a change in the spatial extent of the storm. The model runs also projected that with a 2°C rise in temperature, precipitation in Colorado could increase by as much as 30 to 40 per cent, which is three times more than what would be expected given the Clausius-Clapeyron relation (the principle in atmospheric physics which defines how much water vapour can be transported in a warming or cooling atmosphere). The net effect of more rain, heavier snowfalls at altitude and increased runoff will likely be larger floods. Dr. Rasmussen is now doing a supercomputer model run that could answer one of the most critical questions posed at the workshop: If future precipitation does increase by 35 per cent in Colorado, could flooding of the magnitude experienced in the front ranges of the Rockies in Alberta and Colorado in 2013 become a more regular occurrence? Stay tuned.

Dr. Rasmussen spoke again that evening, at a public session sponsored by the Town of Canmore, about the Colorado front range flood of 2013 and the lessons it holds for Alberta. He began with visuals showing the extent of the inundation and the damage it caused to homes, roads, bridges and other infrastructure. He noted that 262 homes were destroyed in Boulder County alone and that some 1300 landslides were attributed to the flooding. Dr. Rasmussen commented that it was highly unusual for such a storm to occur in September, and that the cause of the flooding was a low-pressure system that parked over Colorado and continued rotating, all the while bringing up tropical moisture from the south. Rasmussen said the storm was remarkable in how much precipitable water it carried in the form of atmospheric water vapour, more than tripling the former record. The storm also set new records for daily, monthly and annual precipitation, despite the severe drought that had preceded it. The previous Colorado record for rainfall in 24 hours, 4.8 inches, was nearly doubled by the 9.08 inches that fell in 2013.

Dr. Rasmussen went on to point out that the Alberta and Colorado floods were quite similar in a number of ways. Both involved slow-moving upper-level low-pressure systems bounded by high-pressure systems to the north and south. Both low-pressure systems remained nearly stationary for an unusual length of time and worked with nearby atmospheric cells to bring large amounts of water vapour up from the south. And both of the resulting storms were of long duration, delivered heavy rainfall and covered a very large area. The differences were that in the case of the Colorado flood there was no snow on the ground and the lightning events were more numerous. Rasmussen noted one other important difference, which had yet to be determined: the Colorado flood may in fact have been caused by what can be defined as a tropical, as opposed to a temperate-region, in-cloud rainfall formation process.

As for lessons Alberta could learn from the Colorado event, Dr. Rasmussen pointed out that accurately predicting heavy precipitation is difficult in all of the US National Weather Service climate models. The service had been working on this problem for some time, and the 2013 flood accelerated research efforts to improve ensemble analysis of coupled hydro-meteorological changes in real time. Quantitative weather radar was invaluable in the US and not normally available yet in Canada; the NowCast and Forecast products are part of a new US system that has emerged since the 2013 storm. There is still work to be done, however, to identify sources of error and opportunities for improvement.

WHAT DID WE LEARN? WHAT DO WE DO NEXT?

So what do we do now? With respect to describing floods and their statistical properties, workshop participants proposed the following. We need to go beyond conventional analysis in characterizing the frequency of flood events. We need to look at more than simple examination of annual extremes and include all events in a single year. In these types of analyses we need to think more about flood mechanisms, including, for example, rain on snow and antecedent moisture conditions and their effect on potential flood intensity. Consideration should be given to how the potential for increased flooding is affected by historical and recent changes in land use and vegetation. We should compare historical flooding events in different basins to look at similarities and spatial correlations. As the climate continues to change, we should investigate how large-scale precipitation events are also changing in terms of magnitude, frequency and duration and look into the role of teleconnections and the transport of moisture northward from the Gulf of Mexico. Geomorphological processes and other dynamics of the systems need more consideration. This includes, for example, sediment redistribution on alluvial fans and changes in stream channels, which affect the hydraulic regime and may lead to reductions in future stability and certainty in behaviour. Groundwater and its effects on persistence of high flows should also be considered. It is important that statistical analyses include confidence ranges for the results, as this is useful for planning and operations and for public awareness. Finally, it is also very important to communicate research findings and our uncertainties more effectively and more widely.

In considering challenges and opportunities related to modelling, there was consensus among workshop participants that we need a broader range of better observations, including precipitation and other meteorological measurements. This would improve data assimilation into models. Other suggestions on how best to model floods included continuous flow simulations, physically based modelling, ensemble forecasting, clear understanding of hydrological processes and model parameter limitations, incorporation of human impacts into models, coupled atmospheric–hydrological modelling, and improved ice-jam modelling. Some other points noted included the importance of publicly disseminating model outputs, and that models can be used to help develop indices for conditions related to flash floods, landslides and debris flows.

The western Canadian flood in June 2013 and the Colorado one the following September make good potential case studies for modelling. For the Canadian flood, a legacy dataset should be developed and a special-issue journal publication should emerge. A modelling intercomparison study in connection with the US National Center for Atmospheric Research, incorporating the western Canadian and Colorado front ranges floods, would be a useful exercise that would provide valuable insight.

The group also noted some longer-term issues and opportunities. These included coupled atmospheric–hydrological modelling and ensemble modelling; risk communication; technology transfer of newer models to the provinces and territories; linking drought and flood modelling through continuous simulation; model intercomparison efforts; and climate and weather change modelling studies. A key point for the modelling perspectives was that, once again, researchers heard that they have to be better at public dissemination of model results.

With respect to water management and flood mitigation, workshop participants proposed that we need to translate research outcomes into useful tools for engineers, architects and planners. For engineers, the needs and implications centre around updated storm intensity, duration and frequency analysis and peak discharges for infrastructure design. For municipalities this involves stage and water levels and updated flood maps, while for water managers this would include reservoir operation. Some challenges for reservoirs were noted. Among these were different philosophies on the purpose and operation of the reservoir as they relate to, for example, drought resilience, water supply, hydro-power generation and flood protection. The current state of water resource systems modelling presents challenges for reservoir operation, where daily and hourly simulation is needed to assess flood management, and flow routing needs to be incorporated into current schemes. Other challenges and important issues canvassed were that uncertainty propagates from observations and models up to the management decision level, making reliable forecasts critical for reservoir management. Scenario assessment and economic and ecological risk and benefit analysis are useful for better evaluation of possible outcomes, and this is something the research community should strive to enhance and to include reservoir operation.

In terms of mitigation, we need enhanced zoning and socio-engineering solutions, improved public awareness and education programs, better risk quantification and communication to the public and to political leaders, and forecasting tools that will help arrive at stricter floodplain zoning. It was also pointed out that floods have important and beneficial environmental functions, such as redistribution of sediments and nutrients, and that these benefits should be considered along with the risks.

After the close of this important Canmore conference, researchers once again came away with great respect for the work that is being done and for the people who are doing it. After the disaster of 2013, they were in even greater awe of the challenges to come. They were also left with the relentless reminder that they have got to get better at sharing, with political leaders and the general public, what they know now and what they will be learning in the future. These are the kinds of promises that are made in the wake of a storm.

CHAPTER SEVEN

FOR WHOM THE BELL TOLLS

Lake Winnipeg and the Prairies
in the Anthropocene

Not everyone in Canada, or anywhere else for that matter, gets the opportunity of timely reconsideration of their hydro-climatic circumstances in advance of rapid change. In some regions, like the Lake Winnipeg basin, hydro-meteorological change is already occurring and appears to be accelerating. In such places, major efforts are required to first catch up with these changes and then, if possible, get ahead of them. The case of Lake Winnipeg is particularly complicated by the fact that it involves the Red River, which of course originates in the United States. It also involves a century-old treaty between the two countries and an organization created under that treaty for the purpose of anticipating and resolving potential disputes over waters shared by the two countries.

A public forum on the Boundary Waters Treaty, the International Joint Commission and the Future of Lake Winnipeg was held at the University of Manitoba in Winnipeg in May of 2014. The sponsoring partner was the Manitoba Institute for Policy Research, a think tank nested within the University of Manitoba's political studies department. The conference was a joint presentation with the Forum for Leadership on Water, or FLOW, and was advertised as part of the institute's "Citizen Series" of presentations aimed at enhancing the quality of political discourse in Manitoba.

The program was opened by Paul Vogt, formerly the top civil servant in the government of Manitoba and now the director of the Manitoba Institute for Policy Research. The event was moderated by

FLOW co-chair Norm Brandson, who, having spent many years as a deputy minister in Manitoba's Environment and Water ministries, knew the nutrient issues related to Lake Winnipeg as well as anyone in the province.

The keynote presentation was by FLOW's Murray Clamen, who until very recently had been secretary of the Canadian section of the International Joint Commission. Dr. Clamen is an expert on transboundary pollution in the Red River basin and related matters along all of the fluid points of the international border. His keynote addressed the delicate political question of whether the IJC should receive a reference mandate to resolve potential disputes related to the effects of transboundary nutrient flows from the US into Canada and into Lake Winnipeg by way of the Red River.

Dr. Clamen began by explaining what the International Joint Commission is; how it came into existence as a consequence of the signing of the Boundary Waters Treaty in 1909; and how the reference mechanism functions to formalize dispute resolution in cases of Treaty violation or potential threats to harmonious relations between the two nations over shared waters. Dr. Clamen summarized the remarkable success the IJC has had over the past century in identifying and addressing issues before they became disputes and in resolving disputes amicably to the satisfaction of both nations. He then explained that the water quality issues underlying recent claims that key conditions of Article IV of the Treaty were being violated were in fact valid, and that an IJC reference was perfectly in order to address what had the potential to create tension between Canada and the US over the deteriorating condition of Lake Winnipeg.

An astute and experienced diplomat, Dr. Clamen left much unsaid. Citing research findings by Dr. Greg McCullough (which we'll return to in a moment) pointing to American sources as the main contributor of phosphorus into Lake Winnipeg, Dr. Clamen left it to his audience to figure out the extent to which these inputs violated the Treaty provision dealing with transboundary pollutants. He also left it to the audience to determine for themselves whether a reference to the IJC was warranted. He did this not to elicit hard feelings between the two nations but to demonstrate that we are not utilizing all the resources that are available to us to address the Lake Winnipeg problem.

This failure, however, goes beyond the issue of Lake Winnipeg. In failing to utilize historic resources that have served both nations well, we are in addition eroding important, long-standing, hard-won avenues connecting our two nations politically which will be difficult to recreate if they are lost to neglect and disuse. Clamen concluded by calmly and articulately presenting the logic behind preventing the terms of the Boundary Waters Treaty from being ignored and then forgotten in our time.

A four-person panel of local and international experts, including this author, then responded to Dr. Clamen's presentation. Panellists made it clear that Manitobans face some serious and mounting problems. The panel noted that the Lake Winnipeg problem was too big and far too complicated to rely on government alone to address. Civil society has to work with government of course, but there is also work to be done within civil society itself to begin addressing the problem. The entire process of collaboration needs to be re-energized.

An open forum followed for questions from the audience. A commenter suggested that all parties need to use all the tools that are available to address the eutrophication problem in the Lake Winnipeg basin. It was also noted that equal and similar rights in Canada's relationship with the United States had been institutionalized through the Boundary Waters Treaty, and that a hundred years later the rest of the world is still trying to catch up with the degree of fairness of equity embodied in the transboundary relations between the two countries.

The IJC, it was noted, is the main instrument that puts the value of the Boundary Waters Treaty into relief. But because we no longer use the IJC to fulfill its original purpose, the Treaty is losing its force. One of the panellists argued that in addition to being an outright violation of Art. IV, the claim North Dakota is making – that the state government should be permitted to unilaterally construct major diversions because in funding these themselves they hold such projects to be exempt from the Boundary Waters Treaty – is legally indefensible and warrants an IJC reference in its own right.

Similarly, major flood storage plans being developed in the US by the Red River Basin Commission should be properly vetted under the Treaty, if only because these plans interfere with the current hydrological regimes in both countries. While everyone on both sides of the border may agree that such plans will serve to reduce the growing flood threat, no accommodations have been agreed upon to determine how and to what extent holding back water in the US might affect Canada during prolonged drought.

At present neither country appears to care much that articles of the Treaty are being violated and threatened with further violation. We don't appear to want to use the IJC in a proactive way prevent and resolve disputes between good neighbours. But if we really do need all the tools available to us, then maybe that is a mistake. Evidence from the Great Lakes suggests we shouldn't abandon approaches that have been successful in the past.

It was noted that IJC studies fundamentally underpinned the Great Lakes Water Quality Agreement. While the engagement of the IJC did not involve a formal reference that would trigger a dispute resolution process, it did serve as an effective platform for co-operation between the two countries in dealing with the shared problem of deteriorating ecosystem health in the Great Lakes. Lake Erie is the closest analogue we have to Lake Winnipeg in Canada. It is roughly the same size and has many of the same problems. We have not been able to adequately sustain successful collaboration in the Great Lakes, and now four of the five lakes are in measurable decline. Erie is threatened once again with death. Is Lake Winnipeg going to suffer the same decline?

As the conference proceedings were taking place at St. John's College on campus, it was not surprising that a church bell tolled each quarter-hour throughout the course of the evening's presentations, panel discussion and final summary. It seemed at those moments that the bell tolled for Lake Winnipeg. As Norm Brandson noted in closing, we don't have sixty years to get on top of the lake's eutrophication problem. And if that is the case, then that bell tolls also for us. If we don't want an IJC reference on phosphorus loads in the Red River, then another agency of its kind has to perform the function of unifying efforts of government at all levels; of harnessing science and interpreting its outcomes; and of re-energizing civil society in both countries in the interest of the common good. The Lake Winnipeg situation will get away on us if we can't organize ourselves to act. We need the equivalent of a Great Lakes Water Quality Agreement for the Lake Winnipeg basin. Can the Red River Basin Commission, the Lake Friendly Accord and Alliance - or a combination of the two - become the IJC equivalent in this process? Or will we just forget the Boundary Waters Treaty and relegate the IJC to other, lesser roles than dispute resolution and proactive diplomacy between the two countries? These questions were not addressed as the bell tolled a final time as the forum came to an end.

A number of participants later gathered at a nearby pub, where the owner, who happened to have a significant interest in a company that manufactures and sells flood protection systems, made a point of reintroducing himself and his company's products. It was an astonishing coincidence that while he did so the bar's very loud sound system was belting out the famous Creedence Clearwater Revival song *Have You Ever Seen the Rain*. In the background you could still hear that bell toll.

LAKE WINNIPEG AND ITS BASIN: HYDRO-METEOROLOGICAL CHANGE AND ITS CONSEQUENCES

The rain *is* coming down and the nexus between water, food, energy and biodiversity is not holding. After the prairie flood of 2011 the Centre for Hydrology at the University of Saskatchewan expressed the view that what was happening in the Lake Winnipeg basin was emblematic of how the speed at which water moves through the hydrological cycle is accelerating in every part of the country. The manner of its moving is changing too. It appeared that the kind of flooding that was occurring was evidence that the central Great Plains were approaching - or perhaps had even passed over - an invisible threshold into a new hydro-climatic state. Particular concern was expressed about the Lake Winnipeg basin, where mobilization of nutrient-rich agricultural runoff during floods was contributing to what potentially could become one of the largest freshwater ecosystem collapses in the world. While Canadians are shocked that such environmental catastrophes are occurring in their country, there are few places in the world where they are not occurring.

We are learning the hard way that hydro-climatic stability is a very valuable common-pool resource without which some economic sectors such as agriculture will find sustainability an ever greater challenge. We know that hydrological conditions on this planet have always been changing, but usually that has happened at a much slower pace. We have been fortunate to have had a century or so of relative hydro-climatic stationarity. That era is over. This is a huge new concept – a societal game changer – and it is going to take time to get our heads around it.

As discussed in chapter five, the loss of hydro-climatic stationarity makes the established bell curve of climate risk meaningless. In a more or less stable hydro-climatic regime you are playing poker with a deck you know and you can bet on risk accordingly. The loss of stationarity is playing poker with a deck in which new cards you have never seen before keep appearing more and more often, ultimately disrupting your hand to such an extent that the game no longer has coherence or meaning and can no longer be played. Unfortunately, the Lake Winnipeg basin is the first heavily populated region in Canada to be affected in this way. In this basin we face a growing number of "no analogue" situations where we are facing conditions we haven't seen before. We are all concerned, for example, about the flooding that occurred in July 2014 in Saskatchewan and Manitoba. In discussions with experts, it seems that what occurred that summer appears to confirm earlier fears that this region has indeed crossed over an invisible threshold into a new and more energetic hydro-climatic regime.

In December 2013, University of Manitoba research Greg McCullough reported that algal blooms on Lake Winnipeg had reached 17,000 square kilometres in area. McCullough was also able to provide the latest research findings with respect to sources of the nutrients that were finding their way into the lake. He reported that there was 10 times the volume of phosphorus in the Red River as compared to the Saskatchewan and Winnipeg river systems. He also noted that the annual flow of the Red has doubled over the past 20 years. As a result of this increase there have been more floods. The consequence is that more water with twice the phosphorus per unit volume is now entering Lake Winnipeg.

The median total phosphorus concentration in the Red River in flood is twice the concentration in normal summer and fall flows. The concentration of phosphorus in the Red increased by 30 per cent between 1971 and 1980 and by 45 per cent between 1996 and 2005. As a result the productivity of Lake Winnipeg and its fishery has grown but there are serious concerns that the expanding extent of eutrophication will result in anoxia and a sudden collapse of the food web in the lake, which would destroy the fishery. No one knows at what threshold that would occur, however.

We see from this that Manitoba faces a much larger problem than just the eutrophication of Lake Winnipeg. While that is a big enough problem in its own right, Manitoba now also faces the likelihood of a significant increase in the frequency, duration and extent of flooding as a consequence of regional land-use alteration, drainage effects, streamflow changes and the intensification of the larger hydrological cycle. This one-two punch will make it harder to bring Lake Winnipeg and waters like it back from the brink of total eutrophication. Increased flooding could also have a self-limiting impact on the region's ultimate agricultural productivity, as well as unexpected impacts that are passed on to others by way of downstream flooding that accelerates eutrophication not just in the Lake Winnipeg basin but in lakes and streams throughout southern Manitoba and the Great Plains.

More than ever we are worried about what the loss of hydro-climatic stability might ultimately mean to the economy and future prosperity of Manitoba and surrounding jurisdictions. The duration of precipitation events on the Canadian prairies has lengthened by half since 1940. Total precipitation has grown by 30 per cent in some parts of the region since the 1990s. The growing length and intensity of precipitation events has already resulted in a fourfold increase in streamflow during such storms. Researchers have demonstrated that the growth of these larger streamflow volumes and runoff ratios since 1994 is likely due to non-linear, threshold-crossing responses to a combination of changes in land use, recent dramatic increases in wetland drainage and a changing climate.

The effects of drainage have recently received a great deal of research attention. A long-term measurement and computer modelling study undertaken by the University of Saskatchewan's Centre for Hydrology has revealed that wetland drainage is a major factor in increased prairie streamflows and greater flooding in wet years. Because of the intensification of agricultural practices, the basin of Smith Creek, Saskatchewan, just south of Yorkton, has undergone substantial changes in drainage patterns. In 1958 there were some 98 square kilometres of wetlands which covered 24 per cent of the basin. Today there are 43 square kilometres of wetlands covering only 11 per cent of the basin. The prairie hydrological model was set up to cover the years from 2007 to 2013, using reliable weather data from a University of Saskatchewan monitoring station. These years included the largest flood then on record, which took place in 2011. The model was manipulated to decrease or increase wetland volume by simulating both drainage and restoration. In the model, the target for restoration was the measured wetland extent in 1958, while the limit for drainage was complete wetland removal. The study found that wetland drainage has a significant impact on streamflow in flood conditions.

For the flood of 2011, complete drainage of the existing wetlands would have raised the flood peak by 78 per cent and increased the streamflow volume by 32 per cent. Restoration of wetlands from their 2011 extent back to their 1958 size would have reduced the 2011 flood peak by 32 per cent and decreased the total volume of streamflow by 29 per cent. What's more, it appears that drainage has an even stronger effect on streamflow in normal years. For those years, streamflow volumes would increase by a factor of 200 to 300 per cent with drainage of current wetlands, and the yearly peak flow would increase from 150 per cent to 350 per cent.

Over the six years covered by the computer model simulation, total streamflow volumes increased by 55 per cent with complete drainage and decreased by 26 per cent with restoration of existing wetlands to their 1958 state. We have already discussed what all this means. It means we may now be experiencing non-linear, step-like changes in the manner in which water moves through the hydrologic cycle. If this is the case – and scientists believe it is – the mobilization of nutrients that cause eutrophication is not all we have to worry about. An entire new runoff regime is coming into existence that could further exacerbate eutrophication while at the same time dramatically increasing flood damage and the risk of flooding's diametric and symbolic opposite, drought.

Perhaps the most frightening finding, however, was that no amount of wetland restoration would be adequate to prevent flooding if extreme weather events become more frequent and severe. To be meaningful, wetland restoration must be tied to effective efforts to reduce soil loss, improvements in erosion control, expansion of vegetative buffers and better nutrient management. Without these measures being taken, eutrophication of lakes and watercourses across the prairies cannot be addressed.

If what happened at the Prairie Improvement Network conference held in Virden, Manitoba, in March 2014 is any indication, farmers are already beginning to figure this out for themselves. It was made clear at that conference that southern Manitoba is where everything used by everyone upstream ends up. This includes all the water they use and everything that is dissolved in or mobilized by that water. If this is creating disaster after disaster in southern Manitoba, how long will it be before these impacts advance upstream? Blaming one another for these circumstances is neither fair nor helpful. We have unfortunately inherited the sum total of a century of trade-offs, and unless we want to abandon parts of southern Manitoba, something has to be done about it.

Three months later, as if to completely underscore the outcomes of the Virden workshop, southeastern Saskatchewan and southwestern Manitoba flooded again. While many considered the flooding in 2014 to be déjà vu all over again, it was different from 2011 in one fundamental way: it was caused not by spring snowmelt but by heavy and prolonged late-June rainfall. Other factors exacerbated the flooding. Heavy winter snow had already saturated prairie soils made even wetter by the heavy spring rains. A huge frontal weather system pushed into Canada from the United States and stalled over southeastern Saskatchewan on June 28. The stalling of these large systems over the prairies, which appears to be occurring in both summer and winter, is being attributed to the loss of Arctic sea ice, which appears to be slowing the northern hemisphere jet stream, making it linger longer over many places, bringing heavy precipitation and more erratic weather to many regions as far south as the mid-latitudes, including the Canadian prairies.

Torrential rains toppled streamflow records set during the 2011 flood. Parts of the region received as much as 200 millimetres over a single weekend – almost as much rain as normally falls in the dry region of southeastern Saskatchewan all year. By July 4 more than a hundred communities in Saskatchewan and downstream in Manitoba had declared a state of emergency. Some 300 people in Saskatchewan and nearly a thousand in Manitoba were forced to evacuate because of overland flooding. Some 8000 in Saskatchewan alone were without power, and the Trans-Canada Highway was closed for days at Brandon. Prime Minister Stephen Harper flew from Ottawa to survey the flood damage.

Manitoba has 160 kilometres of provincial dikes that are almost a century old. No one was sure which ones would hold. The Canadian military was called in to protect 350 rural homes, 150 of which would have been flooded had it been necessary as a last resort for the province to deliberately breach the same dike at Hoop and Holler Bend that had been cut in 2011 to reduce the downstream flood risk on the Assiniboine River. Anger and bitter resignation were registered widely but particularly in communities and on farms that had been flooded by the 2011 dike cutting, who were still disappointed that anticipated compensation and promised action to prevent the same damage from occurring again were still not forthcoming. Worn-out and upset flood victims charged the government with incompetence and pleaded for outside help.

In southwestern Manitoba 50 per cent of the land was covered by water, much of it in the form of floodwaters coming from Saskatchewan. The flooding was so serious that farmers had to ignore the state of their crops and give full attention to saving themselves, their homes and livestock. A million acres of productive prairie farmland went unseeded. We are now finally able to calculate the real costs – including the costs of lost productivity associated with extreme weather. It is now estimated that economic losses due to the Prairie drought of 2000–2004 exceed $4-billion, while the damages from the Alberta–Saskatchewan–Manitoba floods of 2011–2014 top $11-billion.

If issues related to hydro-climatic change were not well understood before the flooding in 2014, it can be said with a high degree of confidence that after what happened that summer, some farmers at least are starting to get the message now. In an example mentioned at the end of that summer by one farmer, hundreds of thousands of tonnes of sediments from Deer Horn Creek were washed in torrents into the Assiniboine River, changing channel morphology in ways that are sure to greatly increase the risk of future flooding. Argued a rural mayor, "It is suicide to ignore the clear and obvious fact that our hydrology is changing." The days of a farmer gaining fifteen or twenty acres of productive land through drainage that causes the loss of four or five times that as a result of flooding downstream have to come to an end. "We have to learn to live with reality," the mayor of Virden said. "We need to learn to deal with this rather than continuing to fight it." Not everyone in Saskatchewan is on the same page with drainage impacts yet, but getting and keeping them onside is critical if only because in the Anthropocene everyone has to be in on the solution or no one will succeed. If current trends persist, it may not be that difficult to convince enough people to act collaboratively to address the problem.

THE IJC RECONSIDERED

It may be useful in the context of the eutrophication issues and loss of hydrologic stability in the Lake Winnipeg basin to examine other regions that have recently faced inadequate or exhausted relations that needed to be renewed with surrounding jurisdictions over shared water resource matters. Fortunately a peerless example exists right here in Canada to guide us to the resolution of these difficult problems.

Probably the best example of success gained and then squandered in terms of the capacity for Canada and the US to come to full and effective agreement on the co-management of shared waters is the Great Lakes Water Quality Agreement of 1972. How Lake Erie nearly died, was saved and is now again under threat demonstrates that indigenous peoples should not be the only ones to be concerned that we as a nation have begun to turn our back on the personal and environmental protections that at one time defined our national character. The concern – as has been noted in the Lake Winnipeg basin – is that we are on slippery slope of ineffective action that may over time put any meaningful economic and environmental sustainability out of reach for everybody. This is not a foregone conclusion, however.

What many experts in Manitoba appear to be aiming for in the Lake Winnipeg basin is an as yet undefined, perfected form of adaptive, basin-scale, integrated water resource management. Four elements have been identified through the evolving practice of integrated water resource management which, theoretically at least, are held to build adaptive capacity in socio-ecological systems. The first is learning to live with change and uncertainty. The second is nurturing the kind of diversity you need to create resilience. The third is combining different kinds of knowledge. And the fourth, the creation of self-organizing pathways leading to socio-ecological sustainability. These are the principles that are driving the UN's *Transforming Our World: The 2030 Agenda for Sustainable Development* as well as the various "resilient cities" and similar sustainability movements around the world.

What we learn from Lake Erie is that we are not starting all over here. We know how to do this because have done it before. Research conducted recently by Dr. Savitri Jetoo at McMaster University demonstrates that the determinants of adaptive governance of the kind we seek today were clearly present 50 years ago during the reversal of the eutrophication in Lake Erie, beginning with the Great Lakes Water Quality Agreement in 1972. We just didn't call it "adaptive governance" back then.

The Great Lakes Water Quality Agreement spurred federal–provincial co-operation in Canada. The agreement worked because it went on to encourage binationalism; promoted community participation; demonstrated equality and parity in its structure and obligation; arrived at common objectives such as joint phosphorus reduction targets; facilitated joint fact-finding and research; demanded accountability and openness in information exchange; and featured flexibility and adaptability to changing circumstances.

Looking back at it now in the context of the reconsideration of the Lake Winnipeg problem, it appears that the greatest success of the Great Lakes Water Quality Agreement was that it fulfilled the promise of institutional integration under a binational framework where common interests superseded political partisanship and jurisdictional territoriality.

There was widespread consensus among key players at the time that the International Joint Commission played a crucial leadership role in all the processes that led to nutrient reduction in Lake Erie. It was the IJC that orchestrated significant public participation and ensured that science was central to the decision-making. It was the IJC that helped create strong action-oriented networks charged with implementing recommendations. And it was the IJC that made sure that information-sharing protocols actually worked and that clear leadership was demonstrated in the creation of a common community that was kept going through reliable, adequate funding.

Looking back, many believe it was the sense of community created by the agreement process that drove the political will of both federal governments to meet the obligations of the 1972 agreement. The 1972 agreement went on to be strengthened in 1978 with the addition of the concept of ecosystem management for the elimination of toxic contaminants. The breakthrough there was that in this process science was meant to guide policy, and public reporting was introduced to strengthen accountability in an expanded mandate that now included air, water and all living organisms in the Great Lakes system. Everything was working.

But then the process started to go backward. After an 11,000 tonne limit on phosphorus was set for Lake Erie, the lake began to recover. It was thought that the problem was solved, so the federal government stopped investing in monitoring and research. Because improvements in water treatment and the banning of phosphorus in detergents appeared to work, changes in agricultural practices remained voluntary. Now Lake Erie is back to the same near-death state as it was in 1972.

Why? What is the difference between then and now?

What is different today is that the population of the Great Lakes Basin has grown dramatically and agricultural production in the region has increased. What's more, the introduction of invasive species has complicated the eutrophication problem, and changing hydro-meteorological conditions have resulted in more frequent flooding, which is mobilizing more nutrients.

What is also different is that the institutional arrangements that made it possible to address the eutrophication of Lake Erie 50 years ago do not exist today. Back then, the federal government was fully funding the Centre for Inland Waters as well as a wide-range of critical research and monitoring programs. Since then, however, monitoring and research have been drastically cut and the remaining federal scientists have been gagged. The technical capacity formerly present in the federal government no longer exists.

Most critically, the political leadership that enabled the saving of Lake Erie in the 1970s is today almost completely absent. We no longer have a reliable mechanism for implementing achieved agreements in a timely manner. Our binational integration processes have been weakened by overrepresentation of provincial and state interests at the expense of partnership and favourable regional and binational relations. This is a trend that has to be reversed if Lake Winnipeg is to be saved.

What we can see by comparing our circumstance then and now is that there is a direct link between the diminished role of the IJC and our failure to uphold today our historic accomplishment with respect to eutrophication in the Great Lakes. It should be remembered that in 1964 the governments of Canada and the United States issued a reference to the IJC with the goal of determining the exact causes of eutrophication in Lake Erie. Funding of the Experimental Lakes Area in Ontario resulted in a broad range of scientific work that changed our understanding of aquatic ecosystem dynamics. That research included some globally significant breakthroughs made by scientists of the calibre of Dr. David Schindler and his associates in the understanding of nutrients.

It should also be noted that although the conventional wisdom today is that the IJC is too bureaucratic and takes too long to render reference decisions, this is not historically accurate. What careful historical research has clearly demonstrated is that delays in handing down IJC decisions were largely the result of the inordinate length of time the two federal governments took to advise the IJC of their wishes. Surveying all of the facts, it is hard not to conclude that political influence and lobbying by special interests have played a huge role in delaying important decisions – so much so, in fact, that it is difficult not to believe that special interest influence did not intentionally seek to undermine the purpose and function of the IJC.

So, why do we need the IJC today? We need it because if we don't have it, the Boundary Waters Treaty becomes meaningless just when we need it most. The eutrophication of Lake Erie in the 1960s was a straightforward challenge compared to what we face today as a result of new, interacting stressors such as invasive species and climate change. More than ever, we need the neutral, non-partisan, binational objectivity invested in the IJC through the Boundary Waters Treaty.

The chair of the US section of the IJC, Lana Pollack, was the keynote speaker at a lunch session of the Red River Basin Commission annual conference in January 2015. Pollack began by recalling that IJC commissioners are not instructed to uphold each country's interests, but to do their job impartially under both flags. Given that 43 per cent of the international border is water, she said, to think and act otherwise is unproductive. The responsibility must be to the shared waters. The IJC is more about political will than science, she said. Science is easier than politics. Getting people in powerful positions to listen and give consideration to science is not as simple as it might appear.

One of the major problems we face, Pollack stated, is that no one wants to give anything back. Once a government, a corporation, a sector like agriculture, or even an individual acquires rights to a given amount of water or the rights to use water of a certain quality for specific purposes or permission to release certain things into that water without penalty, they don't want to give up those rights or, in many cases, even permit modification of them. They will use political channels to fight having to give anything back. The IJC, she said, is fully respectful of these realities.

One of the IJC's principal duties is to report on how the respective governments are doing with regard to meeting the obligations of the Boundary Waters Treaty. Pollack made it clear to what

I thought was a somewhat indifferent if not rather hostile audience that the IJC was never going to regulate them.

Pollack then offered a very thoughtful caveat. Having reminded participants that the IJC was not a regulatory body in the usual sense, she noted that regulations should not be viewed as restraints but as protections. "You are not allowed to pollute," she said. "Is that a restraint or a protection?" She then said what no one else at the conference had been prepared to say, and it took the IJC to say it: we are not getting there, by which she meant that we are not solving the problems we have created for ourselves in the Red River system, a failure symbolized by the deteriorating state of Lake Winnipeg. In other words, we are not meeting the conditions of the Boundary Waters Treaty.

In the context of the Treaty, it is important to note that Lake Winnipeg is not a body of water shared by Canada and the US, and thus it is outside the jurisdiction of the IJC. But the Red River, which flows into Lake Winnipeg, is a transboundary watercourse that does fall within the commission's jurisdiction. The question once again becomes whether we are going to devolve responsibilities for meeting the conditions of the Boundary Waters Treaty to individual states and provinces or restore the function of the IJC. If Manitobans don't want responsibilities for meeting the conditions of the Treaty to fall into the hands of regional interests without national oversight, then they have to call upon their provincial and federal governments to make a request for an IJC reference that would have to go beyond the situation of Lake Winnipeg itself to address the much larger problem Lake Winnipeg only partially symbolizes. That problem is the issue of transboundary standards, not just for nutrients but also for heavy metals and other contaminants for which no current standards exist.

On this matter Lana Pollack very diplomatically pointed out that there are a number of Canadian problems the US would like to dispute, including Alaska's concerns about contaminants flowing into that state in the waters of the Yukon River; the effects of sulphite mining on Lake Superior; and selenium mobilization in the waters of the Columbia as it flows from Canada into the United States. From this it is clear that transboundary issues related to the violation of the Boundary Waters Treaty only start with our concerns about Lake Winnipeg.

In this context, what Manitobans may wish to do is ask the ijc to study nutrients with the aim of assessing what it will take to protect the Red River as part of a larger goal of establishing transboundary standards for nutrients and metals on all shared waters. It could and should be argued that there is urgency in doing so, for without a baseline of knowledge, issues such as these will only get more difficult to manage under greatly energized and far more unpredictable hydro-climatic conditions.

In addition to transboundary standards for nutrients and heavy metals, we need to consider cyanotoxins. Here the spectre of hydro-climatic change looms large. As water temperatures rise and cyanotoxins become harder to control, we will have no choice but to view eutrophication as a public health hazard. There are no standards for cyanotoxins in drinking water in either the US or Canada. If we shift the problem of eutrophication from solely protecting the environment to protecting our own health and regulate it from a public health standpoint, we broaden and strengthen the framework for eutrophication governance. This is adaptive management. For that adaptive management to work, we need the ijc.

In conclusion, there is a great deal we can continue to learn from the challenges the ijc faced with respect to Lake Erie. Solving the eutrophication problem is one thing; keeping it solved is another. Just as happened with Lake Erie, the Lake Winnipeg problem started out as just an issue of excess phosphorus. Although total phosphorus has decreased, the volume of reactive phosphorus from agriculture has increased over the past 40 years. The effects of long-term farm nutrient practices and more frequent intense precipitation events have converged with the introduction of invasive species to bring the eutrophication problem back into existence in Lake Erie. These same factors are already making the problem more complicated in Lake Winnipeg and could make it a lot worse in the future – which will likely alter the political landscape every bit as much as it has affected the region's hydrology.

WHAT WILL AN ANTHROPOCENE FUTURE BRING?

The Anthropocene on the Canadian prairies is going to be a very turbulent time. With further climate disruption virtually inevitable, it won't be long before there are enough people who rely on the prairies and whose lives have been disrupted by extreme weather to constitute a new political force.

A great deal of what Canadians take for granted will be challenged in the Anthropocene. The notion that humans are somehow separate from the rest of nature, and that preserving the non-human areas of the world because they possessed some particular ecological, aesthetic and recreational value would be adequate to preserve hydro-climatic stability, will also come under fire. The idea that there are no limits to human adaptability – and the underlying societal belief that we can either adapt to any circumstance or alter them to suit ourselves through geo-engineering – will be severely tested and very likely rejected. The legitimacy of our economic system will also be highly taxed. Our blind faith in "free markets" will certainly be challenged. The belief in monetary magic, popularized in the 18th century, which holds that economic markets in a capitalist system are somehow "balanced" by the actions of an unseen, immaterial power that ensures that such markets will always function efficiently to optimally address real and enduring human needs may not survive long into the Anthropocene. Unless we change how and what we value, continued market failures will reveal the "invisible hand" of the marketplace to be that of an intergenerational pickpocket.

Those areas of the Earth's surface that used to be frozen, including glaciers, ice sheets, sea ice and permafrost on land, are likely to disappear. Snowpack and the duration of snowcover in much of the northern hemisphere will be reduced. Precipitation patterns on the prairies have already begun to change. As already noted, with each degree of warming, the global atmosphere can carry 7 per cent more water vapour. Projections for the centre of the North American continent are for temperature increases of as much as 5° to 9°C. The basic laws of physics make it likely that floods in the regions will become ever larger. The loss of hydro-climatic stability may make the Lake Winnipeg problem as much as an order of magnitude more difficult to resolve.

It is interesting to note that California's biggest floods have been caused by the storms popularly known as the Pineapple Express. These storms collect huge amounts of water vapour as they cross the Pacific from Hawaii and unload it as heavy rain when the storms hit the coast of North America. When moderate in scale, they bring badly needed water to a dry state. When larger, they cause flooding of a magnitude we can hardly imagine. The atmospheric river that caused epic flooding in California during the winter of 1861–1862, for example, hammered the west coast of North America from Mexico to Canada for 43 days.

The US Geological Survey has published a new emergency preparedness scenario which demonstrates that if struck by an atmospheric river event like the one that happened back then, California would lose one-quarter of all its homes, with projected damage of $725-billion.

The other thing we need to keep in mind is that water vapour is a powerful greenhouse gas in its own right. The more water vapour there is in an energized atmosphere, the warmer it becomes. Obviously, this will become a climate feedback in itself. As we have seen, other critical weather phenomena are also tremendously affected by differences in the temperature gradient between the poles and the tropics.

There was a time when it was thought that the first clear signal the public would pick up as to the extent and rate of change in our climate would be an obvious decline in local ecosystem function. That was wrong. The surprising thing about ecosystems is how resilient they are. Under the influence of changing temperatures, they will move out from under you, leaving in their wake opportunities for invasive species which will succeed or fail as they may until some sort of new ecosystem stability emerges. As long as human numbers and activities don't cause the Earth system to move to a completely new and different state, life on Earth will likely adapt to these changing conditions just as it has for billions of years. We may find ourselves in what we consider a much altered and diminished world, but life will go on with or without us.

What has captured public attention with respect to climate disruption, however, is the increasing threat it is posing to public safety and its growing impact on expensive infrastructure. The cost is enormous and has been rising for 40 years. It is becoming increasingly clear that our failure to break the silence over climate disruption and change the behaviour of our society in ways that lead to action could cost some of us our prosperity. It could also cost some of us our lives. What the loss of hydrologic stability tells us is that true sustainability may be beyond our grasp if we don't do the right thing now. This suggests we have to view sustainable development in a completely new light. Such development as we have defined it in the past is not enough. What we need is restorative development. We cannot simply accommodate ourselves. We have to put vital Earth-system function back in place in doing so.

GETTING AHEAD OF THE KEELING CURVE

Charles David Keeling began precise measurement of the concentration of carbon dioxide in the Earth's atmosphere on behalf of the US National Centre for Atmospheric Research in 1956. As the Keeling Curve so elegantly demonstrates, our atmosphere is in part the inhaled and exhaled breath of life on Earth. The northern hemisphere breathes in with the coming of spring and breathes out in autumn.

It took only two and a half years of observations on Mauna Loa in Hawaii to see what was going on. With each semiannual breath, the Earth system is inhaling and exhaling more carbon dioxide. By the early 1960s the scientific evidence was clear and the implications of rising carbon dioxide concentrations on climate-system function were obvious. We now fully understand how sensitive the global atmosphere is to what we put into it and what the consequences of ignoring changes will be.

Leaders are defined as those who can identify the major trends that will influence their business or organization and ride the crest of those trends toward a sustainable and profitable future. The Keeling Curve may well be the defining trend of our age. If you are in a leadership position and ignore this trend, you may not be there long.

We are faced with the fact that we have no choice but to accept that our climate is no longer stable and that this poses a huge danger to our future. If a sustainable world is what we want, we have to catch up with, get of and stay ahead of the Keeling Curve. So how are we doing in Canada?

According to the eighth annual RBC Canadian Water Attitudes Study, public opinion may be shifting, but only very slowly. Four in ten Canadians still have no idea where their drinking water comes from. This has changed little over the past eight years. The survey also shows we are still complacent about climate change. Few of us know our hydrology is changing. Only 28 per cent are concerned in any way about future water security. So where is the possible shift I mentioned? Here it is: the 2015 survey found that 7 per cent of Canadians were personally affected by flooding in 2014 and 27 per cent knew someone who was. An amazing finding is that some 30 per cent of Canadians now claim to have been impacted in the past by flooding in their communities.

Insurance matters have also begun to haunt Canadians. Some 30 per cent of respondents said adequacy of coverage was an issue for them, and 55 per cent registered serious concern over their ability to pay additional costs from flooding not covered by insurance.

It appears it will just be a matter of time before the growing number of people personally affected by climate disruption forms a constituency in itself. You can bet that if what we have seen elsewhere is any indication, the economic consequences of climate disruption won't be the only concern on the political agenda of this constituency. The moral, ethical and legal implications will also be high on these citizens' to-do list.

It is interesting to juxtapose these results with those of another survey, conducted by the mayors and reeves of the south basin of Lake Winnipeg, into major concerns over the deteriorating condition of the lake. The survey was based on new techniques for analyzing how best to communicate urgency to different types of people in ways that transcend traditional demographic categories such as age, gender, education, ethnicity, income and net worth. The survey revealed that despite the millions of dollars spent on trying to convince Manitobans of the seriousness of the problem they face, only a third of the messages being sent connect with those whose behaviour needs to be influenced if the province is to address what has become one of the largest and most alarming freshwater disasters in the world.

What we learn from this applies nationally. We have to stop demonizing one another. We must adapt our messages to different kinds of people who have different views and values. We need to give people the capacity to understand and think about water and climate on their own terms and then work back from there. Working back from there means reversing the damage we have done to the Earth-system function and doing it promptly, before population growth, more ecological decline and greater climate disruption further destabilize our already fragile global economic system and reverse hard-won development. To restore our global hydro-climatic circumstances to tolerable stability we have to break the socially constructed silence related to climate disruption. Perhaps the best way to start that conversation is to talk about something we can all agree that none of us can do without: water.

If Canadians want to continue to inhabit vulnerable parts of southern Manitoba, we have to keep nutrients and water on the land, not just within the boundaries of Manitoba but everywhere upstream. People in the know throughout the Lake Winnipeg basin have been talking for decades about a basin-scale approach to management of the lands that affect the quality and flow of the valuable water resources of this agriculturally vital region of Canada. If we want agriculture to continue in parts of Saskatchewan and southeastern Manitoba, then we have no choice but to work together toward the objective of effective basin-scale land and water management.

Continuing to manage water in the same jurisdictionally fragmented way as we do today will eventually bankrupt us. There should be no blame attached to arriving at this realization, though. In the Anthropocene, human populations will be concentrated on the as yet undefined and only partially claimed frontier where water, food and energy meet. If anything, the movement toward meaningful integration of land and water management at the basin level should be viewed as the next step in the maturation of the West, a step that would bring the prairie provinces and states closer together and solidify the globally critical agricultural future of the region. Such a step would also heal the growing rift between rural and urban while at the same time allowing the centre of the continent to become resilient enough to hydro-climatic change to sustain its prosperity long into the future.

Hydro-climatic change, however, is not going to be a simple obstacle to overcome. We need to pay attention to the tolling of the bell. The Lake Winnipeg problem will get away on us unless we can accelerate progress toward better governance without engaging in the usual unproductive partisan politics presently associated with this issue. The Anthropocene offers both an opportunity and a reason to do so. The making of a new, more vibrant and resilient West is not going to be easy. As the Northwest Territories example demonstrates, vision is necessary, and inspired leadership will be required at all levels of government and in all economic sectors if success is to be achieved. Such a transformation will not be possible without the support of an informed and relentlessly courageous citizenship. To elicit that support so that we can get at the very complex kinds of challenge posed by Lake Winnipeg and associated hydro-climatic change, we need to reframe the problem.

At present, policy with respect to dealing with these issues in Canada is moving along at 5 kilometres an hour when the problem is moving at 15 kilometres an hour and accelerating. Catching up with the problem demands a new level of consciousness and non-linear problem solving. It requires creating a new steady state by being dynamic but at the same time self-regulating. In the context of Lake Winnipeg and Manitoba's growing flooding problems and drought vulnerability this means making radical improvements in agricultural practices and integrating supply management with water quality objectives and flood and drought protection through enhanced monitoring and prediction on a basin scale. But before we do that we need to agree on what kind of West we would like to create. Catching up with and staying ahead of the Keeling Curve demands that we decide what hydro-climatic steady state we want where we live and then set self-regulation on the road to achieving that state. What is needed most now is a sense of urgency. The bell is tolling. We know what to do. But we have to do it.

Welcome to the Anthropocene.

CHAPTER EIGHT

VISUALIZING A DIFFERENT TOMORROW

The Columbia River Reconsidered

Aboriginal peoples in the Columbia Basin are poised to redefine the nexus between water, food, energy, biodiversity and social justice and equity in North America. In his book *The Comeback*, prominent intellectual John Ralston Saul observes that with the help of the Supreme Court of Canada, indigenous peoples in Canada are at last regaining the respect, dignity and social, political and legal powers they deserve. Saul quotes indigenous leaders who believe that the rest of Canada should consider doing the same.

It is not the first time such provocative ideas have been put forward. At the Columbia Basin Symposium held at Fairmont Hot Springs, BC, in the fall of 2013 – which Saul attended and discusses in his book – indigenous leaders not only demonstrated their newly restored rights and powers but also made a strong and very articulate case for other Canadians waking up to the realization that serious contemporary threats to their fundamental rights also exist and need to be addressed if they do not want to spend generations fighting to regain those rights after losing them as was the case with indigenous peoples.

The purpose of the symposium was to share understanding of holistic watershed governance as a framework for the impending reconsideration of the 50-year-old Columbia River Treaty under which the US and Canada operate dams and reservoirs on the shared waters of the Columbia to provide flood protection and generate electricity. The symposium explored draft principles for an expanded binational Columbia Basin water governance entity, and right from the start it put into relief stark historical contradictions.

On one hand were experts on the Columbia River Treaty who spoke on how each of its clauses had been carefully and clearly negotiated and each of its conditions precisely met over time through well-established channels of co-operation and mutually respectful, proactive conflict avoidance and resolution mechanisms. These are vital elements of any successful treaty, of course. On the other hand were indigenous people who, though their traditional lands were often submerged, were not consulted in the Treaty negotiations. The treaties to which these people were already signatory, if they existed at all, were dishonoured and continually manipulated to the point of injustice, resulting in what the Supreme Court of Canada described as egregious suffering. Today such conduct could be judged a crime against humanity perpetrated against peoples who in good faith had entered into nation-to-nation relations with federal and provincial governments.

The contrasts between these two views of Treaty relations, and the comparative commitment to meeting their terms and the changes that have taken place in our society since the signing of the Treaty, could hardly have been more obvious. It is not just indigenous rights that have been clarified in Canada during the past half-century. A new and growing understanding of ecosystem dynamics centred around the nexus of water, food and energy has also emerged. We now realize that although the Columbia is a big system, it is not just its sheer size that distinguishes it. It is also unique in its combination of latitude, longitude and altitude, the patterns of its rain and snowfall, the nature of its soils, the broad extent of its forests and the coldness and cleanness of its waters.

We now understand far better than we did in 1964 that the evolution of extraordinarily productive ecological circumstances in the Columbia basin are the result of synergy between ocean, atmosphere, land, water and rivers that together, over millions of years, have created what could be described as a veritable perpetual-motion machine of energy and food production.

Evidence of the productivity of this self-regulating, self-sustaining system is made obvious by the largest salmon runs in the world. Estimates of the peak runs in the Columbia system prior to damming ranged from six million to as many as 16 million fish each year. We will never know how many there really were, but we do know what the iconic species symbolizes. As the author of this book noted in an earlier, collaborative work, *The Columbia River Treaty: A Primer*, salmon could be considered as electrons in the current of planetary life. Existing at the nexus of water, energy and food in the Pacific Northwest, salmon are a key species which transform the enormous energy of the ocean–land–life–water–weather–climate cycle, of which the Columbia is such an enormous expression, into bite-sized bits that other life forms can metabolize. This is why, wherever we find salmon, we also find indigenous peoples whose cultural heritage has linked them directly to the ecological energy of the Columbia River for hundreds of generations.

It is this globally significant energy system that was ignored and compromised in order to harness and control water in the Columbia basin. Because we did not have a full understanding of the riches we already had, we took one of the greatest natural ecosystem powerhouses on the planet and shut it down just to produce electricity. The resulting dams caused convulsions throughout the natural energy cycle of the Columbia system. We realize now that we turned off our biodiversity-based life-support system in the Pacific Northwest just so we could turn on some lights.

SALMON, ECOLOGY AND HUMAN SOCIETY IN THE PACIFIC NORTHWEST

In her 2014 book *Salmon: A Scientific Memoir*, Jude Isabella observes that this is a cold-blooded species that cannot adapt to warming waters. As temperatures rise these fish look for cool pools in rivers – what researchers call thermal refuges – where they can rest and save energy. The range of temperatures salmon can survive without suffering heart failure is quite limited. The Chilko race, Isabella reports, can manage up to 22°C, but the salmon of the Nechako will stop swimming once the water reaches 20°C. Temperatures in salmon streams in many places in western North America are already reaching these temperatures, making it a race to understand salmon before they become largely extinct.

There are also temperature-related issues with the metabolism of sockeye fry in lakes. Such fry do best in water at or below 15°C. If the temperature is higher than that, the fry's metabolism accelerates to such an extent that the available food can be barely enough to sustain them. Warmer water also has a huge influence on disease proliferation. It is now thought that a form of salmon leukemia has been responsible for the decline of the Fraser River population over the past twenty years. The average temperature of the Fraser has warmed 2°C in the last ten years. The Yukon River temperature has risen a whopping 6°C over the last decade. Isabella reports that this warming has been accompanied by the rise of the *Ichthyophonus* parasite, which is killing chinook salmon. Nor are artificial fish-rearing programs bringing these populations back. Hatchery success for chinook in the Phillips River, Isabella reports, is less than 1 per cent.

Isabella's book also explores Aboriginal relations to salmon. She describes in exquisite detail how coastal tribes not only survived but thrived despite chaotic shifts in ecosystem function in the North Pacific, through the use of elaborate stone fish weirs. The use of such structures, Isabella notes, predates the Magna Carta and they remain so clearly identifiable archaeologically that their presence is now being used in settling land claims. While such weirs may be impractical today, what they say about ecologically based cultural resilience is not. The thinking behind these weirs may be important to reconsider today. There has been no time in the evolutionary history of our species when it has been more important to incorporate ancient knowledge into sustainable modern practice.

Part of the value of Jude Isabella's book resides in the fact that it traces the history of salmon, coastal ecologies and salmon-dependent cultures through the upheaval that has been European settlement of North America and into our very uncertain times. Although she doesn't mention it, this idea has considerable bearing on the impending reconsideration of the Columbia River Treaty. Salmon streams are much richer in nutrients - nitrogen in particular - than streams that do not support salmon. By blocking salmon, the Grand Coulee and other dams on the Columbia rob the upper basin of nutrients, which diminishes upstream ecosystem vitality. The dams also terminated the long history of co-evolution of salmon and humans in the Columbia basin. Warming temperatures now call into question whether or not salmon could exist in the future even if the Columbia dams weren't there. That said, we know very little about salmon while they are in the ocean.

We are now using science to try to find our way back to where we were before we transplanted ourselves out of nature, to see where we might somehow fit in again. We are beginning to sort out what Europeans thought they saw when they arrived in the Pacific Northwest versus what was actually there. As Isabella explains, we are relearning the meaning of old words like taboo, which can refer to places you don't fish because that is where stocks replenish. Not only did arriving Europeans misread the meaning and function of ecological processes that animated the Pacific Northwest, they also completely misunderstood the cultural traditions that had emerged from hundreds of generations of close contact with those processes.

The indigenous peoples of the Northwest Coast had a completely different idea of wealth than the European settlers that overwhelmed them. To Aboriginal peoples wealth was simply the fact of living in a place and sharing local relations in such a way as to allow nature to generously supply abundant, delicious food and satisfy the material and spiritual needs of all for all time. Ceremonies such as the potlatch were high-profile expressions of the equitable sharing of that wealth and the commitment societies never to allow material possessions to stand in the way of the common good.

Isabella invites us to give new consideration to our environmental and cultural misreading of the Pacific Northwest. She urges us to go back to that point in our own relationship where we took the wrong turn in our understanding of where and how we should live in such a place. We may in fact wish to go so far as to seek to be neo-indigenous in order to recover harmony and sustainability in our relations with the land, the sea and one another. While only part of the abundant natural wealth that existed in the Pacific Northwest at the time of European contact remains, a reconsideration of what salmon mean to us would be a good place to start in contemporary society's last chance to again become native to where we live. Jude Isabella doesn't think it is too late. She believes that after six million years on Earth and 12,000 years migrating in and out of the Pacific

Ocean, salmon must be at least as able to adapt to climate change as we humans believe we are. Isabella also has confidence in science, which she holds to be the equivalent of a modern potlatch in the degree to which it is generous with its wealth, altruistic in its aims and in constant unselfish pursuit of truth.

Scientific ideals are not as incompatible with traditional values as we used to think. In concluding her book, Jude Isabella makes a most eloquent claim. We need to stitch back together again the larger narrative of the Pacific Northwest, a tapestry of which salmon are just a part. The vehicle for starting that long but necessary process leading to a potentially highly satisfying regrounding in place is the reconsideration of the Columbia River Treaty.

Such matters remain difficult to discuss in the Columbia basin, though, partly because discourse about the Treaty and the future of the basin as a whole has been dominated by those with special interests who still think it wisest to define the future by the narrowly technical and economic parameters established in 1964. The Columbia Basin Symposium put into relief just how easily Canadians have come to accept the notion that the way things were done as a matter of course in 1964 somehow provides valid historical justification for perpetuating denial and injustice today. But just as we begin to think we can carry on exactly as we have done for the past 50 years, the old maxim that what goes around comes around enters stage left in a feather headdress, singing to the beat of a rawhide drum.

TREATIES, SCIENTIFIC TRUTH AND THE HONOUR OF NATIONS

As Ojibway grand chief John Kelly predicted in 1977, what Canadians did to the Indians is what we are now doing to ourselves. We are stepping back and away from honour and truth.

In the same manner as we reneged on our promises as a nation in making treaties with indigenous peoples, we are now backing away from values and commitments that once defined Canada. One of the most clearly obvious ways we are backpedalling away from truth is how we are compromising the bridge we have been building for more than two centuries between science and public policy. Science is to our culture what traditional knowledge is to indigenous peoples. It is the ground of experience based on cumulative observation upon which we base decisions that will affect our future. It is our intellectual potlatch.

This bridge between objective knowledge and evidence and public policy in Canada has come under siege in recent years. We have been muzzling scientists and eliminating science from public discourse. We have cut public support for independent research and delivered science into the self-interested hands of the private sector. In so doing we have allowed the bridge between research and fact-based policy-making to become a fortified border, a checkpoint through which only the science approved by special interests is permitted to pass.

Because indigenous people have experienced this themselves and know how hard it is to regain lost rights, they rise up in outrage with respect to these matters. A comfortable, indifferent settler populace, however, has appeared perfectly satisfied to sit back and watch while their fundamental rights and protections were taken away one omnibus bill at a time by a government that publicly pronounced that obtaining a majority meant it could do anything it wanted as long as it was in power. History, however, has a way of stumbling backwards and colliding with itself. Perhaps not surprisingly, all of these problems have come to a head in the reconsideration of the Columbia River Treaty.

The questions regarding the Treaty have become these: Are we going to carry on saying this was just how things were done back then and keep denying that what was happening then is still happening now? Are we going to continue to ignore Supreme Court of Canada rulings on indigenous rights? Are we going to say it's okay that the same exclusionist interests that crafted the Columbia River Treaty are going to be permitted to perpetuate its terms for another fifty years? Are we going to fall for the worn-out conventional wisdom that Canada – or anyone – could never get a better deal anyway? Are we so comfortable as a society that we will just sit back and let whatever happens happen? Maybe, but not likely. Why? Because the indigenous peoples in the Columbia Basin are not going to let the injustice to which they were subjected in 1964 happen to them again. And if we were smart, the rest of us in Canada would realize we don't want that to happen to us either.

It is now widely recognized that there is a great deal more at stake in the Columbia River Treaty renewal negotiations than is presently understood by the general public in either country. Times have changed. Really changed. Not only are there new ways of estimating the economic value of environmental services, but there are also higher ideals to which we know we can aspire when it comes to transboundary agreements over water. At stake are ongoing efforts to restore river ecosystem function that has been damaged or lost as a result of dam construction. We now realize that sustainable development means restorative development. A primary goal of a renewed Columbia River Treaty should be restoration of lost ecological elements and conditions that fifty years later are seen to be of far greater importance than the architects of the original treaty were able to imagine.

 The second thing a renewed treaty could – and should – do is redeem injustices that today would be judged as human rights violations: actions that took place as a consequence of the non-inclusive manner in which the treaty was negotiated and the final conditions that were imposed upon those who were made subject to its terms.

Also relevant to the Treaty reconsideration is the ability of the people of the Pacific Northwest to monitor and respond to hydro-climatic change in the larger Columbia Basin in a way that will adequately permit coordination of effective and meaningful adaptation strategies. A renewed Treaty should lay the foundation for continuous improvement of social and economic resilience in the face of direct and indirect effects of climate change, not just in the basin but also in surrounding regions. In this context, flood control amid changing hydro-climatic circumstances will be more valuable than ever. So will drought management. And yes, we need electricity. Sustaining productive irrigation is important too. But salmon and other critical ecological values also count.

From this we see that what is really in play here is nothing less than the region's real prosperity, however you wish to define that concept. But there is also more at issue that goes far beyond the river basin. It is clear to many observers that the Columbia River Treaty has the opportunity to become the first transboundary water agreement in the world to be effectively reformed so as to create a living blueprint for how people would like to inhabit their river basin – and regions like it – now and in a sustainable future.

What is at stake, finally, is our moral duty. The reconsideration of the Columbia River Treaty is an opportunity to show the world how to shed the limitations of the past in ways that will allow others to use our example to break out of the prisons of treaties that no longer respond to the realities they face and the new challenges that are emerging as the global hydrological cycle responds to a rapidly warming atmosphere. As Yakama tribal elder Gerald Lewis pointed out at a conference on the Treaty in Ellensburg, Washington, in 2012, we have to keep in mind that we are not reconsidering the Columbia River Treaty just to satisfy ourselves. We are doing this for future generations.

For our children's sake we cannot let what happened in 1964 happen again – to any of us. We can no longer ignore the existence and aspirations of our indigenous neighbours. We cannot afford to allow what was done then to be done yet again to the landscapes upon which all of us have come to depend for our identity and ultimately for the sustainability of our livelihoods and prosperity. We have to heal the waters.

SO WHERE DOES THE PACIFIC NORTHWEST GO NOW?

The first thing we have to do is recognize that the Columbia River Treaty is only one element of a larger whole that needs to change. Reconsideration of this Treaty, however, could be an important door-opener for reconciliation with the peoples with whom we are grateful to share the waters of the Columbia basin and to whom we owe such a considerable debt for having never lost their intergenerational connection with the circumstances and cycles through which human existence derives its deepest meaning and connection to place – a place that is now our place also.

The Columbia basin is not the first place in Canada to confront such matters. The government of the Northwest Territories and its indigenous government partners have faced up to exactly the same kind of challenges and have succeeded in redefining the nexus between water, food, energy and biodiversity while at the same time reconciling equity and social justice in one of the largest river basins the world: the Peace–Athabasca–Mackenzie system. There is much that can be learned from the NWT experience that could help the people of the Columbia basin achieve a better future through reconsideration of this Treaty. Upon examination we discover there is actually a great deal of similarity between the two situations.

Not unlike the residents of the Columbia basin, the people of the Northwest Territories were worried about climate change effects. What is important about their example is that they acted on those concerns. As noted in chapter three, the territorial government and its indigenous, federal and provincial government partners have demonstrated that there is nothing in the Canadian federal–provincial political structure that makes it impossible to undertake the kinds of reform necessary to adapt successfully to climate change through enlightened management of water.

The NWT example does demonstrate, however, that it is not possible to craft leading-edge transboundary agreements with riparian neighbours without first getting your own water policy house in order. Without the *Northern Voices, Northern Waters* stewardship strategy in place, the Northwest Territories would not have had the administrative capacity, scientific wherewithal or moral ground to stand on in negotiating the kind of progressive transboundary agreement it signed with Alberta in 2015.

So what should British Columbia and the residents of the Columbia River basin take away from the Northwest Territories example? The national and international significance of the *Northern Voices, Northern Waters* strategy document derives from the patient and inclusive manner in which it was crafted and implemented and the ongoing engagement through which its effectiveness will continue to be thoughtfully monitored and evaluated. In that it respects traditional knowledge and ways of life and balances the economic realities of resource development within the context of basin-wide aquatic ecosystem health, the strategy represents a landmark in integrated watershed management. Its example demonstrates that in fact there are no real jurisdictional, legislative, constitutional or political obstacles to the creation of a sustainable future based on new ways of occupying the nexus between water, food, energy and bio-diversity. The Northwest Territories example makes it impossible to say anywhere that this cannot be done. If it can be done in the Mackenzie system, it can be done in the Columbia basin. What it will take, however, is collaboration, strong political will and persistent leadership.

As John Ralston Saul reported in *The Comeback*, the indigenous speakers at the Columbia Basin symposium in 2013 told those assembled that the road is open for reconciliation, for correcting the mistakes of the past and for collaboration on a better, more secure, more just and ultimately far more sustainable future for all. They said that this collaboration can begin with water, with the Columbia. Should we not – all of us – take our neighbours up on their kind and generous offer? Is there any reason we can't start again down that road – and that river, together – now?

CHAPTER NINE

FIRE AND WATER

Establishing a Roadmap for Adapting to Climate Change Effects on Water

The Okanagan Basin is a semi-arid valley in south-central British Columbia that has a growing population, significant agricultural development and areas of stressed water resources. Statistics Canada identifies the Okanagan as having the smallest per capita freshwater availability in Canada.

I have to admit I always arrive in the Okanagan with a mixture of delight and trepidation. It is a beautiful place and hospitality there is unfailingly warm and generous. I consider the Okanagan Basin Water Board to be one of the best organizations of its kind in the country and I am always excited to see what they are doing.

My trepidation resides in this region's utter vulnerability to the climate disruptions that are beginning to characterize a new normal in the rest of the country and continent. Every time I go there, the Okanagan reminds me of so many other places where I have observed first-hand what appears to me to be a lethal combination of factors that could very quickly add up to hydro-climatic disaster.

When I land here by plane, I think of the Middle East and in particular Israel with its dry hillsides, dusty mountaintop forests, and its precious valley lake kept, with enormous sacrifice, at a level that sustains its tourism appeal. I am reminded also of the American Southwest, and in particular Phoenix, Arizona, where in the midst of the severest aridity anywhere in the US we find the country's most profligate water use.

Despite the fact that the Okanagan is the driest inhabited region in Canada, per capita water use is around 675 litres per person per day, more than twice the national average and six times the per capita use in water-conscious cities elsewhere in the world.

I think too of California: its exploding population, with climate warming now superimposed over a newly revealed history of century-long droughts and mega-floods, and how the state's powerful economy can suddenly, within a single year, seize up for want of water.

I also worry about the Okanagan in the context of what we are beginning to see happen in the rest of Canada. It appears that changes in hydro-climatic conditions are beginning to cascade through every ecosystem in the country. Recent research demonstrates that if this trend continues, the Canadian West is going to be a very different place. This same research also makes it clear we are not going to have to wait until the end of the century for a new and very different West to come into existence. It is happening now.

When I see what is happening in the interior of the continent, I cannot see how sooner or later these changes will not somehow affect the Okanagan in a profound way. I shudder to think of the region's vulnerability. I worry that it won't be long before the relative stability the people who live there have enjoyed may be disrupted. In saying this I do not wish to imply that this is the end of the world. They know what they need to do in the Okanagan. But what is happening elsewhere in Canada and the rest of the world should add urgency to their efforts. The people of the Okanagan have already experienced disaster. They know what can happen when the rain fails to fall.

OKANAGAN BURNING

It was August of 2003 and there had been no rain in the valley for 44 days. The weather had been hot, with daytime temperatures of over 30°C for weeks. Winter snowpacks, lower than average to begin with, had long since disappeared. This was the driest summer since 1899 and there were no signs the drought was about to end. Rapidly growing populations and the need to keep the level of Okanagan Lake at acceptable levels for the region's critical tourism economy had put huge pressure on already limited water supplies. Rationing had been imposed in the cities, but conservation was not saving enough water to meet all needs. Deep and persistent drought was breaking down water governance structures. Various jurisdictions – from irrigators in rural districts right up to the provincial and federal governments – began arguing over which users were legally entitled to water first. Trout Creek was not the only dispute over water that summer, but what happened there put climate-related water governance challenges in this basin and elsewhere in Canada into stark relief.

Trout Creek enters Okanagan Lake near Summerland, BC. As the second-largest tributary to the lake, Trout Creek provides domestic and agricultural water supply as well as supporting important sport fish – notably rainbow trout and kokanee. In 2003, water storage and intake diversions for irrigation resulted in greatly reduced streamflow. As the drought progressed, less and less water remained in Trout Creek. When ordered by Fisheries and Oceans Canada to restore flows to levels that would protect aquatic ecosystem health, the municipality of Summerland refused to do so, arguing that human needs had to be met first. But it wasn't just the shortage of water that had everyone on edge. Water exists in close relation to its symbolic and diametric opposite, fire. Apprehension about the fire hazard in surrounding forests had been growing for good reason. Without rain, neighbouring timber had completely dried out.

On August 16 the worst fears of emergency services experts were realized. What some have called one of the greatest natural disasters in the history of British Columbia was sparked by a typical summer lightning storm in the middle of the night. The fire started in Okanagan Mountain Park, a 10,000-hectare wilderness that separates Kelowna from Penticton. It was so hot and dry that when lightning struck a single tree it instantly candled into a fireball. Before the resulting conflagration was brought under control, 25,000 hectares of forest had been reduced to ash. Canadian Forces were called in, and 30,000 people – a third of Kelowna – were forced to flee their homes in what became the second-largest evacuation in Canadian history. Though no lives were lost, 250 homes burned to the ground. Damage to built structures and public infrastructure was in the hundreds of millions of dollars.

After the smoke cleared, the residents of the Okanagan rebuilt, but with new respect for what the relationship between climate and water might mean to their future. What the Okanagan Basin Water Board learned for the future is that, under some climate models, the conditions that occurred in Europe and in parts of North America in 2003 could on average be presenting themselves every second year by 2030. Models suggested that similar conditions might arise in one of every four years in the Okanagan by 2018. The board discovered also that they could not adapt to such circumstances under current water governance structures and institutional arrangements.

New ways of managing water are necessary if communities in the Okanagan basin are to become sustainable in the face of a warming climate in what is already the driest inhabited region of Canada. With the help of a wide range of provincial and federal agencies and many local partners, the Okanagan Basin Water Board has started down the difficult road toward policy reform as a means of adapting to climate change. Ten years later, however, the members of the board are still worried that accelerating climate change effects may cause the road they are on to disappear from beneath them before they reach their goal. Evidence from around the world suggests this is not an unreasonable fear.

THE OKANAGAN IN THE ANTHROPOCENE: SOME POLICY RESPONSES

There are very knowledgeable people all around the globe who consider that the hydro-climatic change we are seeing is just one manifestation of much broader effects which human numbers and activities are having on the Earth system. As was noted in earlier chapters, it is being put forward in enlightened scientific and policy circles, not just abroad but also here in Canada, that our impacts are of such a scale that we have, in essence, entered a new geological era in which human activities rival the processes of nature itself. The proposal is to call this new geological era the Anthropocene.

The foundation of the Anthropocene rests on new understanding of how the Earth system actually works. Four components in tandem appear to regulate the stability that makes conditions for life on this planet possible. These include the atmosphere: the air we breathe; the hydrosphere: the water we drink; the cryosphere: the refrigerating influence of the Earth's snow, glaciers and ice sheets; and the biosphere: the living elements of our planet.

Through our emissions we have changed the composition of the atmosphere, and our land use changes and water demands have altered the hydrosphere. Rising sea-surface temperatures are shrinking the cryosphere, and our agriculture and expanding global presence are reducing biodiversity. As a consequence of these compound impacts our climate has begun to change.

From a practical point of view, the idea of the Anthropocene is not as frightening or illogical as it might at first appear. What the concept recognizes is that humanity has in itself become a major force in changing the nature of the Earth's surface. It is hard to deny this. We are veritable chimneys of emissions. At 40,000 parts per billion, our exhaled breath contains a hundred times the concentration of carbon dioxide found in the atmosphere. Seven billion people and their livestock generate seven billion tonnes of carbon dioxide in addition to the 30 billion tonnes we generate burning fossil fuels. Our collective and cumulative breath keeps the world warm. In addition we have become a force in our own right in terms of weathering and erosion. We have also altered global carbon, nitrogen, phosphorus and water cycles. We are causing changes in the chemistry, salinity and temperature of our oceans and the composition of our atmosphere. In tandem our effects are now significant enough to alter weather and climate patterns globally. The concept of the Anthropocene is interesting in that it calmly and dispassionately recognizes that we have changed the terms and conditions of life on Earth. No one intended that humanity should have such an impact on Earth. The Anthropocene concept simply suggests that we should recognize those impacts and manage them.

While many may not agree, some are suggesting that in a way it is a relief to enter the Anthropocene. The idea of naming this new era is that we should stop ignoring what we have done and are doing and start recognizing what we have become. We are now a force of nature, and as such we need to act accordingly in the interests of our long-term survival. We have no choice now, for example, but to work with a changing climate, not against it. We have to prepare for deeper and more persistent droughts and more powerful storms and more frequent flooding, because that is what we are likely going to get. We can already guess what it is going to be like on the prairies. The Anthropocene there – as we saw in chapter seven– is likely to be a very turbulent era.

If there were to be an iconic image of the Okanagan in the Anthropocene, what might that look like? While the Okanagan has greatly benefited from climate disruption to the extent that warming winter temperatures have made the wine industry successful, attention must be paid to changes that are occurring all around us. Researcher Scott Smith has projected that rising temperatures are expected to double the rate of evaporation in the valley bottom. These faster evaporation rates are expected to increase the fire hazard in the Okanagan basin even if precipitation increases.

Like the rest of the country, the Okanagan should also expect more frequent and intense weather events. Flooding, landslides and debris torrents will become more common. Disasters may be triggered more by extreme weather events than by changes in overall weather patterns.

Preserving prosperity in the Okanagan in the Anthropocene will demand that those who live there decide collectively what hydrological and climatic steady state they want and then set self-regulation on the road to achieving those conditions. Some of the best people on this continent are working on these problems. What may be needed now, however, is a greater public sense of political urgency.

No one knows better than the Okanagan Basin Water Board that the watershed basin is the minimum unit at which water must managed. This fact in itself – that basin-scale water management is critical to social, economic and environmental resilience amid changing hydro-climatic conditions throughout the Okanagan – will hopefully inspire greater local action.

The Okanagan is responding to the need for effective water management and has made significant progress in preparing for the anticipated impacts of climate change and population growth with the development of adaptations for water management. Through continuous research since 2000, which that has included climate trend analysis, general circulation model application, and hydrologic modelling and analysis, scientists have been able to identify a broad range of expected influences of climate change on water availability in the Okanagan basin. These include a decline in lake inflows; a change in streamflow timing, with earlier onset of seasonal peak streamflow and an extended low-flow period; increased frequency of drought and/or longer drought periods; and an increase in agricultural water demand due to a longer, drier, hotter growing season. Other projected effects include increases in residential water demand during the growing season, accompanied by additional increases due to population growth; more frequent late-summer water shortages; high variability, both seasonal and annual, in both water supply and water demand; and increases in health-related issues associated with water quality.

For water resource managers in the Okanagan basin, the evolution of climate change adaptation measures has been influenced by a regional policy framework. Within this framework, a decade of collaborative discussions and research, combined with the need for adaptations to the existing governance structure, has led to and legitimized the actions proposed in the Okanagan Sustainable Water Strategy (osws). The osws outlines forty-five actions dealing with water source protection, security of supplies, and delivering on the implementation strategy. The strategy represents a coordinated approach that will promote regional adaptation to climate change as the actions are collectively applied. Currently the actions are at the pre-implementation to implementation stage.

Key factors or concepts that have influenced the successful development of water management adaptations in the Okanagan include a highly inclusive, multi-level governance structure; a strong focus on multi-stakeholder involvement; reliance on local knowledge as well as scientific research; and a regional view of what successful adaptation means. The importance of effective action has also been stressed.

COULD WATER POLICY NATIONALLY EMULATE OKANAGAN PROGRESS?

The aim of this book is to explore the relationship between climate change adaptation and water governance in Canada. Progress in places like the Okanagan notwithstanding, Canada is largely unprepared for climate change. A survey conducted in 2014 by the Federation of Canadian Municipalities revealed that 48 per cent of the jurisdictions that responded did not have formal climate change adaptation plans and were not considering developing any in the immediate future.

Preliminary analysis of the effectiveness of water governance structures in widely different geographic regions suggests that overwhelming obstacles stand in the way of urgently required action on climate change adaptation in Canada. These barriers include the fragmentation of water governance under the Canadian Constitution; reactive governance and lack of a nationally binding policy framework; and information gaps that limit wise management of water in a changing climate. Put simply, water management in Canada is barely coping with the effects of climate change so far, and focused attention is required now if we are to properly address adaptation needs as climate effects accelerate over the longer term. The research that resulted in this book also revealed nested opportunities for addressing the need to change the way we manage water in this country. What is called for is a new road map for making the necessary changes to water governance.

A new road map is imperative because water policy in many parts of Canada has not kept pace with changing political, economic and climatic conditions. While the established apportionment of responsibilities and jurisdiction with respect to the management of water has served our country adequately for more than a century, the policy vehicles that currently dictate how water is allocated and used, and how its quality is protected, were built for an earlier and very different time when the population was smaller and more dispersed and we had fewer competing uses for our water supplies. Our current policy frameworks were also developed during a period of relative climate stability that appears to be coming to an end.

Because water policy in Canada has not kept up with population and economic growth, we find ourselves confronted today with two converging challenges, both of which are being exacerbated by a changing climate. Much of our legislation was crafted to meet

19th century water allocation needs in the context of 20th century technology and hydrological understanding. Unfortunately, however, the way we currently manage water in Canada is not adequate to 21st century circumstances, in which the integration of nature's own need for water and the broader ecological protection of water sources have become recognized factors in sustaining economic and social development.

Because of the way jurisdiction over water is atomized in Canada, it is difficult to change our management practices quickly enough to address the growing array of surface and groundwater problems that are appearing across the country, or to take full advantage of new technologies and best practices that could greatly improve the efficiency of water use in both economic and environmental terms.

Both of these challenges are affected by a changing climate. Climate effects on water have been clearly identified by researchers all over the world. Warming temperatures have already begun to alter the extent and timing of precipitation and runoff in some regions of the country. As was so amply demonstrated in the Okanagan in 2003, we are also beginning to understand the direct link that exists between water and wildfire. The effects of climate change are expected to accelerate and compound in not entirely predictable or uniform ways as global temperatures continue to warm, with potentially alarming effects on water security in some areas. This is not something we can ignore.

In places like the Okanagan, climate change has already demonstrated that our outmoded systems of water management are becoming increasingly vulnerable to disruption, with potentially disastrous future impacts on our economy and way of life. Even in relatively water-abundant areas such as Nova Scotia and the Northwest Territories, climate effects on the delicate balance of water supply to natural systems are of growing concern.

On the positive side, many municipalities, regional districts, some provinces and territories and a number of federal agencies have committed to formal processes of evaluation of adaptation potential at various scales in different regions of the country. Much of the focus on adaptation to climate change has been on ensuring reliable water supply, not just for domestic and industrial needs but also for economic and environmental sustainability now and under possibly changed conditions in the future.

Municipal leaders and policy-makers across the country are discovering that improved water management for its own sake would achieve a number of important sustainability goals in a broad range of economic sectors, which in themselves will add up to greater adaptive capacity. Provided that climate change is contained within reasonable limits by fossil fuel emissions reductions, the harmonizing and streamlining of federal, provincial, regional and municipal water governance structures and functions will go a long way toward make adaptation to climate effects on water resources possible even in already water-stressed areas of the country. At the time of this writing, however, there is little evidence to suggest that fossil fuel emissions in Canada are being reduced in any meaningful way, and harmonization of federal, provincial, regional and municipal water governance structures is happening so slowly that it is difficult to know whether progress is in fact occurring. While some gains are clearly being made in the important area of water conservation, it appears that whatever governance advancements are occurring are still being mocked and overshadowed by population growth and further economic expansion. We can't keep up with the problems we are creating for ourselves. As this appears to be a problem everywhere in the world, there is much we can learn from others.

LEARNING FROM THE GLOBAL WATER PICTURE

Most studies cite population growth as the principal driver of increases in the global demand for water. Although there are always numerous uncertainties surrounding future projections, research shows that the world population is likely to grow by 30 per cent between 2000 and 2025, and by 50 per cent between 2000 and 2050. At a minimum, the fact that global population is expected to grow perhaps as high as 10 billion by 2050 invites questions as to whether there will be enough water to support increases of this magnitude. The concern becomes more urgent when it becomes clear that nearly all of this growth will occur in developing countries. Unfortunately, many of these countries had inadequate or barely adequate water supplies to support the populations that existed even at the turn of the millennium, and they will face correspondingly greater difficulty as those populations continue to swell. It must also be recognized that further economic development is likely to fuel increased demand for water, both directly in the growth of water-consuming industries, and indirectly in the form of dietary and other lifestyle changes which tend to require more and more water.

In short, the global picture appears to be almost inescapably one of growing demand in the face of static or shrinking supplies. Water experts like Steven Solomon have argued that a new world order is about to emerge out of the collision between population and economic growth and our planet's rapidly changing hydrology. Solomon points out that the divide is expanding globally between freshwater haves and have-nots. Many analysts believe this explosive disparity is likely to widen across the entire 21st century political, social and economic landscape. We should expect global water and related food security to be among the most pressing political issues of our time. In this there is opportunity for Canada to prosper economically by helping others.

Provided Canadian agriculture can address its own sustainability problems, we should anticipate that this sector will play a significant role in addressing the issue of food security. But water availability will not just affect food production. The amount of water available to a given nation will determine its industrial capacity and the quality of life its citizens enjoy as a result of nature receiving the water it needs to make places be worth living in. Prosperous countries in the future will be those that have enough water for food, cities, industry and nature – and those that know how to ensure that each of these users gets the water it needs when it needs it.

Canada could be a leader in this emerging new world water order, but only if it can ensure that the serious problems and unfortunate paralyses that have emerged so widely elsewhere in the world with respect to the management of water do not gain a permanent foothold in this country. Although the same population growth and economic pressures that together have created a global water crisis elsewhere have already begun to appear here, Canada has started, albeit slowly, to move in the direction of governance changes which, if they were more focused, could put us in a leadership position. But we will have to move faster if we want to be a player in this new world water order. Contrary to the myth built up over centuries of limitless clean and abundant Canadian water, the fact is that there is no longer any foundation for our national pride in the quality of the water in our lakes, streams and rivers.

While this book focuses largely on climate impacts on freshwater and how governance might adapt to them, climate change is only one of many threats to water security as identified by Environment Canada. Additional hazards include an expanding number of water-borne pathogens; the growing effects of persistent organic pollutants and mercury; the widespread contamination of streams, rivers and lakes throughout southern Canada by agricultural runoff; aquatic acidification; the ecosystem effects of genetically modified organisms; urban runoff and municipal wastewater effluents; industrial point-source discharges; the poisonous effects of mine water, landfills and waste disposal contamination; and growing concerns over the concentration of endocrine-altering and other troubling substances in our water supplies. Together these problems contribute to reduced ecosystem function and further loss of biodiversity. It could be said with considerable justification that we are diminishing our way of life in Canada through ineffective governance of our precious water resources. While that may not be apparent to most city-dwelling Canadians now, our failures and shortsightedness will be put into clear relief as our climate warms.

SO HOW BAD WILL IT BE?

Climate change is a risk multiplier. Canada's water supplies are already under increasing strain, and climate change will not only compound these problems but create new ones as well. Our current, fragmented approaches to water management stand in the way of addressing water management challenges in an integrated way. This in itself is a feedback, if you will, that will likely increase our vulnerability to the effects of climate change.

Changes in Canada's climate are already affecting water systems across the country, with many more impacts anticipated as these changes accelerate and intensify. Climate change influences the hydrologic cycle, which in turn affects water in all of its forms and every one of its biogeophysical functions. Warm the world even slightly, and surface water and groundwater, ice and snow, ecology and habitat, weather patterns and ocean dynamics are all simultaneously altered in response. Experts predict that climate change will increase precipitation, evaporation, water temperatures and hydrological variability across Canada, all of which will negatively affect the quality of our water. While climate change and climate influences on water show general consistency across the country, regional differences are also apparent. In the Yukon and coastal British Columbia, for example, potential changes include increased spring flood risks and impacts on river flows caused by glacier retreat and disappearance that are likely to reduce hydroelectric potential. There will also be ecological impacts related especially to fisheries, as well as damage to infrastructure, and issues over apportionment of water for human uses.

The Rocky Mountains may see higher-altitude snowlines in winter and spring, possible increases in snowfall in some areas and more frequent rain-on-snow events, which will increase the risk of both avalanches and floods. There may also be a decrease in summer and late-season streamflow, which could have ecological effects as well as impacts on tourism and recreation.

On the prairies the possible changes are expected to be manifold. These include changes in annual and summer streamflow which will have implications for agriculture, hydroelectric generation, water allocation and ecosystem health. The increased probability of severe drought has the potential to increase aridity in semi-arid zones, further adding to losses in agricultural production and resulting in land use changes. The combination of these factors is likely to affect surface and groundwater quality and quantity in ways that may have adverse impacts on farm incomes. As noted earlier, a trend toward more intense and frequent storms and floods has already been confirmed.

In the densely populated Great Lakes basin, climate change and climate influences on water are likely to result in precipitation increases coupled with greater evaporation leading to reduced runoff and declines in lake levels. This will have potential consequences for hydroelectric generation, shoreline infrastructure, shipping and recreation. Decreased lake-ice extent, including some years without ice cover, will result in ecological changes and increased water loss through evaporation, which will affect navigation.

In Atlantic Canada, less snowfall and a shortened duration of snowcover are likely to result in smaller spring floods and lower summer streamflow. Changes in the magnitude and timing of freeze-up and breakup will have implications for spring flooding and coastal erosion. Further saline intrusion into precious coastal aquifers is also expected to occur, which will result in loss of potable water and increased conflict over water allocation.

In Arctic and sub-Arctic regions, climate change and climate influences on water are likely to result in thinner sea ice, a one- to three-month lengthening of the ice-free season and an increased extent of open water, which, while improving navigation, will have ecological effects with consequences for traditional ways of life. Terrestrial impacts will be pronounced as well, including loss of permafrost, which will affect ecosystems, wildlife migration, road transportation and human life in many northern communities. As we saw in chapter three, the thawing of permafrost and warming of the Arctic Ocean could also result in a rapid and significant increase in methane release into the atmosphere that could further accelerate climate change, not just in the circumpolar North but around the world.

HOW WELL ARE WE ADAPTING?

The sheer extent of the projected impacts of climate change on water resources across Canada should serve as a strong impetus to ensure that we adapt appropriately to minimize negative effects. Yet our current approach to water management is proving largely ineffective at enhancing our resilience. A June 2010 report from the National Round Table on the Environment and the Economy confirmed that Canada's water management approach is outdated, and highlighted climate change as one of the four most important water sustainability issues affecting the nation, predicting that it will transform the way we manage water. Nothing has changed since 2010.

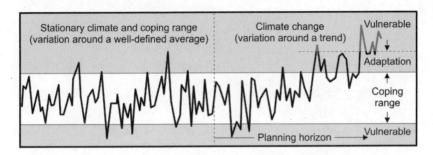

The white strip indicates the fixed envelope of relative certainty within which we have come to expect our weather and climate to fluctuate over a typical period of time. This bounded range of possibility is something engineers, for example, would take into consideration in designing major infrastructure such as bridges or highways. Also shown are the increasing number of extremes we will have to adapt to over time. This loss of what is called "stationarity" is making our society increasingly vulnerable to accelerating climate change effects. One of Canada's leading experts on climate change adaptation, Dr. Ian Burton, coined the term "adaptation deficit" to characterize this failure to recognize or fully acknowledge the evidence that our climate system and its related services are moving in a direction that is beyond the range of our current capacity to adapt.

La Niña years, in which cooler temperatures and greater precipitation occur, will allow some people in some places to keep believing that nothing is changing, but such years will become less frequent over time. Researchers at NASA concluded in January 2010 that despite large year-to-year fluctuations associated with the El Niño-La Niña cycle of tropical ocean temperature, the global temperature continued to rise rapidly in the past decade. In the wake of the worst fire season in history in Russia in 2010, the world reached a record high global temperature as measured by instrument data. Pakistan experienced the fourth-highest temperature ever recorded, 53.7°C, or about 128.6°F, in the spring of 2010. In 2015 drought conditions persisted in Pakistan despite extraordinary flooding in the intervening years.

Ever more frequent extreme weather events continue to drive home the realization that the enormous cost of repairing the damage caused, particularly by flooding, could make it very difficult to sustain our prosperity while at the same time protecting and improving our environment. We should expect increased costs resulting from droughts such as in the Okanagan and floods like Manitoba's.

As those costs rise we should also expect the potential for general tensions and conflict to rise with them.

WE ARE COPING, NOT ADAPTING

The primary response to climate change thus far has focused principally on mitigating it by reducing greenhouse gas emissions. While such action is crucial, it is inadequate by itself. Current and projected atmospheric concentrations of greenhouse gases are substantial enough to mean that further climate change will occur, and indeed is already occurring, regardless of our success in reducing emissions. Therefore it is important to couple our efforts to *mitigate* the cause of the problem – greenhouse gas emissions – with efforts to *adapt* to the current and anticipated effects of climate change. Despite groundbreaking work being done in places like BC's Okanagan basin and by a number of municipal governments, even the most progressive jurisdictions are only in the pre-implementation or early implementation stages of effective adaptation.

Adaptation is intended to reduce vulnerability and enhance resilience. The latter term is defined by the Intergovernmental Panel on Climate Change as "the ability of a social or ecological system to absorb disturbances while retaining the same basic structure and ways of functioning, the capacity for self-organization and the capacity to adapt to stress and change."

In 2009 the Adaptation to Climate Change Team at Simon Fraser University investigated and confirmed the effects of climate change on biodiversity. In his summary report on climate change adaptation and biodiversity, lead policy author Jon O'Riordan recommended that governments transition to a regime in which all land- and water-based decisions are either made by a single agency or coordinated across several agencies as overseen by a single agency, and legislation and regulations be aligned to ensure resource decisions are based on ecological principles that support ecosystem resiliency in a changing climate. British Columbia adopted this approach with the creation in 2010 of the Ministry for Natural Resource Operations, which consolidated all resource decision-making into a single, integrated model. The government is currently analyzing how water decisions will be made under a proposed Water Sustainability Act, which will balance efficiency of decision-making with sustaining the ecological health of watersheds.

This approach, however, is at odds with the linear system of engineering which was the main paradigm governing water management in the 20th century and continues to predominate. In linear system engineering, physical projects are designed to supply water, control floods, improve drainage and support transportation. Although subject to environmental assessments, these projects have tended to degrade ecological system function and require large fiscal and engineering resources. In an adaptive paradigm, the emphasis would switch to "engineered ecology," in which maintaining or restoring ecological function is paramount to retaining ecosystem resiliency. Examples include "low impact design" in the management of rainwater to increase the use of natural drainage systems rather than engineered solutions; demand management and water conservation to reduce the need for new water supply projects; and protecting wetlands and flows in rivers and streams to sustain ecological function.

Such strategies underscore the value of adaptive rather than reactive approaches to managing water. A key strategy is to embed the concept of "public trust" in water governance. This means that all levels of government have a fiduciary duty to protect public rights to clean and secure water supplies and to maintain healthy ecosystems, and that these interests override private rights through licensing and water allocation.

Ralph Pentland is one of the most respected water policy analysts in Canada and a senior member of the Forum for Leadership on Water. As he has pointed out,

> Public trust law or its equivalent goes back to the time of the Romans or even earlier. It is founded on the reality that certain natural resources – especially air, fresh water and oceans – are central to our very existence. For that reason, governments must exercise a continuing fiduciary duty to sustain those resources for the use and enjoyment of the entire population, not just the privileged.

Pentland notes that since the 1970s, a rich body of public trust law has developed in the United States. He cites Joseph Sax, who, in an important 1969 journal article, put forth three key ideas in support of the doctrine of public trust:

- that certain interests are so intrinsically important to every citizen that their free availability tends to mark the society as one of citizens rather than of serfs;
- that certain interests are so particularly the gifts of nature's bounty that they ought to be reserved for the whole of the populace; and
- that certain uses have a peculiarly public nature that makes their adaptation to private use inappropriate.

The principle of public trust began to be undermined in Canada as provincial governments, led by Alberta, made a subtle but ultimately very significant shift in their thinking about environmental protection. The change is best understood by examining how the mandate of many provincial environment ministries changed from "environmental protection" to "environmental assurance." Whereas "environmental protection" meant that landscapes and public resources such as air and freshwater needed clearly defined protection, the notion of "environmental assurance" meant that governments could reserve the right to compromise the level of such protection in the service of economic interests, provided that certain impacts were mitigated and established pollution standards were not exceeded. Environment ministries no longer protect the environment so much as they employ mitigation measures in an attempt to limit the amount of impact economic development may have on it. Their job is to make development acceptable.

After this change gradually occurred, however, most provincial environment ministries found themselves with reduced budgets and diminished capacity, especially with respect to monitoring and enforcement. As industry self-regulation became more common, governments moved further and further away from effective administration of public trust with respect to air and water. The public trust is now seen to have been so violated by resource activity in some provinces that governments as represented by environment ministries are being publicly accused of outright failure of duty and care. Can this be so?

In a 2009 conference presentation called "Fixing Canada's Failing Water Contract," Pentland observed that modern public trust doctrine in the US is primarily a creature of state courts, and that there are in fact as many variants of it as there are states. The real power of the doctrine, he said, lies not in the laws themselves, which are often weak or unclear, but in the creativity of the courts and those arguing cases before them, adding:

> Differences in the nature and scope of public property rights in Canada and the US make an exact duplication impossible in this country. But there is definitely a "trusting relationship between governments and citizens which could facilitate the development of something akin to public trust law in this country.

Pentland has noted elsewhere that:

> there is a lot we could learn from other federated states. The most important thing we could learn from our closest neighbour, the United States, which is also a federation, is the importance of governance systems which are much more open and transparent than ours, which better prioritize the public interest and environmental protection, which guarantee citizen participation in all water management decisions, and which give citizens the legal right to insist that their governments meet their fiduciary duties.

Public trust is the heart of an adaptive approach to water management that builds natural ecological resilience by increasing the capacity of a given water system to absorb disturbances, making it more responsive to change. It requires a focus on maintaining ecological integrity of watersheds; using the same water more than once; an experimental approach that promotes learning; management at the basin scale; multi-level governance; stakeholder participation; open, coordinated sharing of learning; strong and persistent leadership; and the creation of financial and insurance mechanisms that promote appropriate attitudes toward water use and protection.

The most commonly recommended adaptation options for the water resources sector all represent "no-regrets" policy choices, meaning that implementation of them would lead to benefits irrespective of the effects of climate change. These include water conservation measures; providing ecological flows in watersheds; retaining properly functioning ecosystems; improved planning and preparedness for droughts and severe floods; improved water quality protection from cultural, industrial and human wastes; enhanced monitoring efforts; and improved procedures for equitable allocation of water.

Well-planned, practical policy options are essential if we want to effect lasting, sustainable change in our water governance in Canada. But even the most progressive water management cannot succeed in creating the level of adaptation required without breaking down jurisdictional and other barriers to improved watershed governance.

BARRIERS TO EFFECTIVE ADAPTATION

In a 2011 report on the need to adapt to climate change effects on water in Canada, the Simon Fraser Adaptation to Climate Change Team identified several key obstacles to achieving this goal. These include the high cost of infrastructure replacement, jurisdictional fragmentation, reactive governance, policy gaps and gaps in information. Foremost of these is the high cost of infrastructure replacement.

Adaptation to the water-related effects of climate change will require expensive infrastructure upgrades. The cost of such works – and the fact that the Canadian public has swallowed the neoliberal fable that the fiscal capacity of governments to replace infrastructure is limited – are enormous obstacles to increasing resilience and therefore pose significant barriers to effective adaptation.

The second issue, jurisdictional fragmentation, arises because responsibility for water resource management in Canada is diffuse. The Canadian Constitution apportions legislative power over freshwater between the federal government and the provinces, producing a complex regulatory web that spans municipal, regional, provincial and federal levels of government. Although no powers are explicitly delegated for "water" or the "environment," the Constitution identifies many responsibilities that necessarily include these topics. The federal government has constitutional power over fisheries, transboundary waters and First Nations lands; provincial governments have jurisdiction over water quality regulation, allocation rights and land use; and municipalities are most often responsible for land-use zoning, water services and infrastructure.

Furthermore, different ministries and departments have their own internal requirements, standards and regulations, which are often not well coordinated. The intersections of these responsibilities, both vertically and horizontally, create competing pressures that allow key issues to slip through cracks, and throw up confusing barriers for policy practitioners and politicians alike. Many of the same kinds of problems presently exist within and among Aboriginal governments.

The complexity, fragmentation and lack of coordination in water governance in Canada results in policies that are often inconsistent with one another as to drinking water quality, ecosystem protection, allocation rights and climate change adaptation. According to Dr. Karen Bakker, editor of *Eau Canada: The Future of Canada's Water*, the trend of "passing the buck" between levels of government creates "an ill-coordinated downshifting of responsibilities, leaving key areas in a policy vacuum." As Canada's water resources come under increasing pressure from climate change in the coming decades, exacerbating existing challenges and creating new ones, it is crucial that all orders of government address this source of vulnerability by working toward interjurisdictional harmonization of water regulation and movement toward the integration of land and water management at the watershed scale. This is a lot more easily said than done.

The third area of concern relates to reactive government and policy gaps. Due to jurisdictional fragmentation, Canada lacks a clear national vision for managing our water resources, as evidenced in part by the fact that the last federal water policy was tabled in Parliament over two decades ago and has never been fully implemented.

Water in Canada tends to be governed in a less than proactive way, as we have seen in the oil sands of Alberta, for example, and in the case of the large-scale toxic algal blooms in Lake Winnipeg. This failure to act proactively with respect to water governance has left Canada with wide-open gaps in policy, particularly in terms of water quality regulations, where there are major disparities in public safety across the country. As a result of inadequate governance and funding cutbacks in water management, Canada is one of the very few developed countries that does not have legally enforceable water quality standards.

Finally, there are information gaps that exist as a result of incomplete or inconsistent hydrometric monitoring, which makes it very difficult to establish baselines against which we can compare what has happened in the past with what we are experiencing today. Throughout southern Canada it seems we have simply given up on this need for representative long-term monitoring. What we are doing is rather like declaring gauges irrelevant and just taking the dashboard out of your car so you don't know how fast you are going or how much gas you have, and then justifying those actions by claiming you can simply adapt to the consequences. It appears we had a better sense of what we had to know to manage water effectively 40 years ago than we do now. At least then we understood that we couldn't manage what we couldn't measure.

The failure to adequately monitor what is happening to our water resources is held in many circles to be one of the most problematic of all barriers to adapting to climate effects on our water supplies. The fact that both federal and provincial governments have cut monitoring and research related to their own hydrological circumstances could thus be viewed as a failure of public trust, if not a complete failure of government in matters of duty and care.

WHAT WE NEED IS A POLICY ROADMAP

Though there is growing public concern over water security, there has been no change in prevailing rates of water use. In the minds of most Canadians water remains an abundant resource, a free good, an inexhaustible commodity that does not need to be conserved. There is a need in Canada to shift to a conservation society. If such a change does not occur quickly, water security in many parts of the country will be compromised. Climate change will result in more frequent extreme weather events that will cause flooding, more property damage, higher insurance costs and a greater infrastructure maintenance and replacement deficit nationally. Water quality will continue to deteriorate widely. Conditions on First Nations reserves will be even more difficult to address than they are today. Wetland loss and other climate-related impacts will combine with human activities to result in reduced ecosystem services, which will in turn cause lower flows for fisheries, more erosion, reduced groundwater recharge, worsened problems with nutrient and pesticide loading and still more trouble with toxic algae, adding further to water treatment costs. These impacts will be compounded in direct relationship with climate effects and will cost more and more over time, reducing our quality of life and diminishing Canada's economic competitiveness globally.

Neither have governments properly responded to what to some appears a contradiction between the fundamental human right to water, and appropriate pricing of water supply and sanitation services. Governments at different levels are responsible for promoting the fundamental right to basic water supply and sanitation while at the same time introducing appropriate pricing for water services. It cannot be one or the other. Sustainable water management demands both.

Changes in governance to date at the provincial level tend in the right direction for the most part, at least in terms of intention. In practice, however, they fail. In order to meet the growing challenge posed to water security by current and projected climate effects in many parts of Canada, governance has to become more focused and more actively coordinated across the provinces. This can be achieved through the Council of the Federation, which recently approved a new Water Charter and has already committed to working with all provinces and territories and with related organizations such as the Canadian Council of Ministers of the Environment to build watershed stewardship capacity throughout the country. The Council of the Federation has also established a role for expert advisers from outside the public sector to contribute views on new water governance approaches. This development respects one of the advantages of Canadian federalism in that it demonstrates that experimentation at a variety of levels can produce effective, coordinated regional responses to national issues. This approach could be the bridge upon which to build a truly nationwide approach to water policy reform that could in turn become the foundation for successful climate change adaptation, at least as it relates to the management of vital water supplies and the ecosystems that both provide and rely upon those supplies for their integrity and sustainability.

Our changing climate and hydrology require that we shift out of the coping zone of what hydrologists call stationarity and adapt to the new normal. The extent of this adaptation will require that a new set of values underlie water governance in Canada in the future. Fundamental among these values will be the need to balance the water needs of nature with those of people. The creation of such a water ethic in Canada can be achieved in 12 steps. These could form the roadmap we need to finally achieve the water reforms we set out to put in place in 1987.

Step One

Value water appropriately and promote wise use and conservation of it through the establishment of national water conservation guidelines. This step would require provinces, territories, First Nations, municipalities, business and real estate associations, the wider Canadian agricultural community and the energy sector to work with the federal government to set national and regional water conservation guidelines that recognize the value of water. Such guidelines would include building-code standards for smart metering of water-using devices in new homes, with similar standards to radiate outward into all economic sectors. This step would require that conservation be identified as fundamental to a new Canadian ethic, and that cities and regions would compete to be "water smart." Provinces would emulate Ontario in expediting development of clean technologies by encouraging water conservation and water-smart engineering through the creation of tax and other incentives and through the support of independent research clusters that incubate innovation.

Also in this first step, the federal government would encourage home audits of water conservation similar to energy audits, and explore implementation of variable water pricing schedules that do not discriminate against the poor, such as increasing block pricing and seasonal pricing. Also provided would be education so that the public becomes aware of water conservation opportunities in their own homes and businesses and appreciates how water savings are an important practical measure in our national effort to adapt to climate-related impacts on water security.

These actions would also include nationwide regimes to strategically link agricultural policy to water governance so as to reduce farm impacts on water quantity and quality. Such measures would also connect water policy with food security in Canada and abroad as well as with energy policy in order to reduce the impacts of energy production on water quantity and quality.

Step Two

Urge governments to value water to meet nature's needs and ensure that its use is consistent with sustaining resilient and functioning ecological systems. In this step all provinces, territories and First Nations governments would be required to adopt legislation in accordance with agreed standards for establishing, codifying and protecting ecological flows for nature now and in a climate-altered future. Each jurisdiction would create programs aimed at increasing the resilience of aquatic ecosystems to adapt to a changing climate through the establishment of legally defined ecological sustainability guidelines and through investment in ecosystem restoration within established watershed management frameworks. In this step, legislative mechanisms would come into existence to ensure that the value of ecosystem goods and services and environmental externalities such as pollution are accurately, uniformly and transparently accounted for in all development decisions.

Step Three

Recognize and value established knowledge and experience in prediction by strengthening and harmonizing flood protection strategies nationally. In this crucial step it will be necessary to develop a joint provincial–federal–continental program for integrated flood prediction, prevention and management and to strengthen municipal emergency planning to deal with the new normal in natural variation of water cycles. It will be necessary at the same time to establish formal processes to reduce the projected economic costs of flooding, droughts and climate-related extreme events by protection of intact natural systems.

Step Four

Support the design and sustainability of water supply and waste disposal infrastructure based on ecological principles and adaptation to a changing climate, with special attention to First Nations. In order to adapt effectively to climate change effects on its water supplies, Canada needs to set new standards for the design and function of water-related infrastructure. This step recognizes that the effects of increased hydrological variability can be managed through the application of the principles of engineered ecology. In order to pay for these changes in infrastructure design it will be necessary to seek revenues through water metering, equitable water service pricing and technological innovations that include reuse of treated water, energy derived perhaps from fermentation of treated wastewater, and the integration of diverse untapped energy sources such as solid organic residues: food, wood and agricultural waste, lawn clippings, biomass. In this step the integration of rainwater management with wastewater handling would be encouraged as a means of reducing the costs of treatment and the energy required to transport water and treat sewage.

This step will demand the full engagement of the professional engineering and water management community in innovative infrastructure redesign that encourages the regeneration of the ecological health of watersheds through the principles of engineered ecology. This will provide the means and the technological solutions that will allow Canada to address critical challenges for access to clean and secure water supplies in numerous First Nations.

Step Five

Recognize the value of comprehensive monitoring and fulfill the need for accessible information required to manage water in a changing climate. As has been reaffirmed again and again in this book, it is impossible to manage water effectively – especially in a changing climate – without careful monitoring. The importance of a rigorous national framework of consistent and regular hydrologic and water quality monitoring has already been established at Environment Canada through the National Water Survey and its provincial and territorial partners. This framework, however, needs to be incorporated into a permanent national database of hydrological information linked to similar systems in the United States so as to enhance the weather forecasting and climate prediction capacity of both countries.

In order to build on what we have, this fifth step will require the establishment of and permanent support for the monitoring of health effects of changing water cycles nationally, but especially on First Nations reserves, as well as the strengthening of public reporting of monitoring programs. It will also require consistent funding of ongoing science that will analyze the broadest possible range of water issues associated with a changing climate, with a special focus on continued support for centres of excellence in science, particularly as they relate to climate monitoring and water-related adaptations to climate change.

Step Six

Recognize, value and support the role of education in public understanding of the importance of water to our way of life. Lack of public understanding of the seriousness of the water governance issues across the country is one of the major obstacles to the rapid development of adaptive capacity in Canada. In order to change this, the provinces and territories may together decide that a nationwide public education and social media campaign is required to dispel the myth of limitless abundance of water in Canada and to establish and advance the principles of a new water ethic. Such a campaign could serve to reshape public understanding of the importance of water by countering this erroneous notion of inexhaustible supply and explaining the global water crisis and what it means in terms of both challenges and opportunities. In so doing, such a program would assist in the transition to adaptation by outlining changes in habits and other measures Canadians need to undertake in order to assure water quality and security in the face of growing populations, increased economic activity and climate change. Such a program could also outline the extent of jurisdictional and legislative changes that will be required to achieve the ideals of the new Canadian water ethic, and perhaps even more importantly, lay the foundation for public support of political leadership that will reform and harmonize water policy at all levels of jurisdiction over the coming decades.

Step Seven

Recognize water as a human right integral to personal security and health. How we decide to manage water now will cast a long shadow forward. In order to assure a sustainable and prosperous future for those who follow us, it is crucial that we put into place today an ethical framework for appreciating the critical role of water in our lives and our world, a framework that will serve the next generation, who will inherit responsibility for their own children. Step seven encourages a nationwide commitment to honouring the principle of public trust, which recognizes the value of water as a public resource for the betterment and enjoyment of present and future generations.

On one level this means honouring the responsibility principles inherent in Canada's founding water ethic, which is to say the one established by Aboriginal peoples. This means honouring indigenous rights to clean water and sanitation as a matter of equity and justice and as a means to enhance adaptive capacity in the face of climate-related effects on water security. On a broader level it means developing and enforcing national drinking water standards and considering a public interest doctrine for water. It also means discouraging the creation of water markets until full analyses have been undertaken with respect to how current water quality concerns can be addressed and how these and other matters will be dealt with amid changing hydrological regimes in the future.

Step Eight

Provide support for holistic approaches to managing watersheds through collaborative governance. After 25 years with virtually no movement at all, the logjam that has been water policy reform in Canada is beginning to shift. The first evidence of movement presented itself in the form of the new Water Charter put forward by the Council of the Federation, which is comprised of the premiers of all of Canada's provinces and territories. Step eight builds on their work.

This step recommends that the Water Charter be strengthened to include principles for water security as defined by the need to create a new Canadian water ethic. It proposes that the Council of the Federation advance water policy reform in Canada by aligning responsibilities set out for the federal, provincial, municipal and First Nations governments so that they work co-operatively to achieve the principles of climate-related adaptation and the Water Charter. This suggests the need to undertake watershed assessments of entire basins and not just portions of watersheds, as part of development proposals. It also means including water management as a central element in all land-use planning that projects climate-related effects forward to at least 2050.

Step Nine

Recognize the importance of groundwater and urge governments at all levels to understand and value its role in creating a sustainable future. Comprehensive and coordinated national groundwater protection and management strategies should be developed to address current inadequacies in groundwater mapping and monitoring.

Step Ten

Recognize the value of developing coordinated long-term national strategies for sustainably managing water in the face of climate change. The evidence obtained through the research efforts of networks funded by the Canadian Foundation for Climate and Atmospheric Science, as discussed above, will likely give readers a sense of why active adaptation to emerging climate effects on water should demand our full attention. What this tenth step recommends is that the provinces should work together with the Canadian Federation of Municipalities and the most senior levels of the federal government to establish a firmer climate adaptation approach to water management across the nation.

It will be important in this step to identify tangible changes in key management approaches to water conservation, flood protection and emergency measures and improved monitoring and science. It will also be important to develop new water allocation models based on priorities and having the flexibility to deal with droughts and attendant water shortages.

To ensure that real change occurs, it will be crucial at this step to further identify jurisdictional, institutional and legal obstacles to water policy reform and harmonization, and to undertake policy reform and harmonization at local, regional, provincial, territorial and federal levels in order to improve the way we manage water so as to ensure sustainability and security of supply and quality in the face of a changing climate.

Step Eleven

Consider the value of creating a non-statutory National Water Commission to champion the new Canadian water ethic. The Council of the Federation may see fit to become or create such a body, which would undertake a thorough, continuous examination of the kinds of legal, legislative and policy reforms that will be necessary to ensure that water everywhere in Canada can be managed in a manner consistent with integrated land and watershed management principles rather than through the fractured jurisdiction of artificial management units imposed by political boundaries. It will be crucial to undertake this examination in the context not just of the hydrological circumstances of today, but also of the realities that might come into play unexpectedly as a result of climate change. After identifying and prioritizing specific legal, legislative and policy barriers to holistic water management, a National Water Commission might then request that the Council of the Federation facilitate negotiations between all relevant levels of government and with affected economic sectors to gradually but persistently bring about necessary reform.

Step Twelve

This is the ongoing commitment to articulation and promotion of a new Canadian water ethic. As mentioned earlier, the jurisdictional schema of the Constitution would necessitate a multifaceted governance model for water management. By definition, no single level of government can implement policies to improve water security. The Council of the Federation issued its Water Charter in August 2010 as an initial step in establishing a number of water stewardship principles that would bind all four levels of government. This is an important step toward water policy reform in Canada. But it is just a first step and it needs to be followed by many others that have not been taken in the five years since the Water Charter.

The Council of the Federation may wish also to consider the next phase of its Water Charter. In advancing the recommendations above, the Council would build on that Charter to articulate a new water ethic that would redefine the way Canadians value and relate to water. The values inherent in that ethic should include recognition of the crucial need for comprehensive and continuous monitoring of the state of our waters; recognition of water as a human right integral to human health; and recognition and respect for the water rights of indigenous peoples. Such an ethic should also include recognition of nature's own need for water; acknowledgement of the need to break down jurisdictional barriers to more holistic management of water; and recognition of the need for governments at all levels to further advance appropriate economic signals that effect positive change with respect to how water and water infrastructure are valued and managed.

Finally, the adoption of such an ethic must be founded on the wide recognition that successful adaptation to climate change will be only one of many benefits our society will derive from better management of our most precious natural resource.

FROM WATER POLICY REFORM TO BROADER ENVIRONMENTAL AND ECONOMIC CONCERNS

In the course of researching this book, the author visited dozens of Canadian communities. While each was at a different stage of adaptation, the various communities – as widely separated as Kelowna, BC; Sydney, NS; and Yellowknife, NWT – had each arrived at a common conclusion about how to most effectively begin adapting to climate effects that are already beginning to appear. Expert observers in each of these communities stated that the best way of quickly defusing the climate threat is to manage water more effectively and in a more integrated way. They have seen that better water management in itself will make it possible to solve a number of other problems while at the same time assuring that climate-related water quality and quantity issues don't threaten their economic and social future.

This point was particularly well made by a former Nova Scotia deputy minister of environment and labour, Bill Lahey, at a forum held in Cape Breton as part of this book project. Mr. Lahey indicated that a new Canadian water ethic might be what is needed to bring about widespread change in the way we manage and protect water, not just in Nova Scotia but everywhere in the country. Effective policy change, he explained, needs an organizing principle or ethic around which the people and the politicians that represent them can rally. By way of demonstrating the power of such an ethic, Mr. Lahey cited the healthcare debate that took place in Canada during the 1960s, when the ethic of universal access emerged as the driving principle in the reform of our country's healthcare system. He noted that, though simple, the ethic at the heart of this principle was strong enough to survive repeated attacks, even by powerful interests advocating for private healthcare.

Mr. Lahey suggested that climate change adaptation in Canada could be rapidly advanced if Canadians got behind a new water ethic in the same way. He further argued that, if that were to happen, political leaders in many parts of Canada could – because there is less debate about water than many other resources – move quickly to address a good many related environment problems simultaneously en route to better water management. The affirmation of a new water ethic, he pointed out, could be a means of ultimately achieving greater adaptive capacity to climate change while generating numerous other lasting social, economic and environmental benefits along the way.

FAST FORWARD TO THE OKANAGAN IN 2035. WHAT HAS BEEN ACCOMPLISHED?

The drought and fires of 2003 were a crisis, to be sure, but they brought the Okanagan valley's vulnerability into sharp relief and illuminated the need for change. A movement began in which community leaders worked across political boundaries to create a policy vehicle to collaborate on water management. Created in 1970, the Okanagan Basin Water Board had never been fully tested in its role of encouraging local governments to pool resources and make joint water management decisions. After 2003, however, it was given a new mandate to take the lead on regional science and policy for climate change adaptation, population growth and other shifts in society and the environment.

In the Okanagan the first steps taken were to build partnerships and strengthen communication with stakeholder groups, research institutions and various levels of government. The Okanagan Water Stewardship Council was established as a formal way to bring these groups together to advise the Water Board and create the Okanagan Sustainable Water Strategy. The council and other, more ad hoc partnerships have paved the way for basic water science studies and for rebuilding monitoring networks and calculating water needs for fish, food production and basic sanitation. Plans built on good science are more durable and they level the table for working on agreements with mutual benefits. Good planning has begun to shape the form of development and the design of infrastructure – the bones of communities.

This is a new era for the Okanagan. The people in this region have learned that adaptation or "sustainability" isn't a state that a region just arrives at. It is a process of responding flexibly with timely action to situations as they arise – armed with good information. Collaboration, especially for water management, allows for that flexibility. There remains a great deal of work to be done, however. As noted at the beginning of this chapter, per capita water use in cities like Kelowna is around 675 litres per person per day, more than twice the national average and six times the per capita use in water-conscious cities elsewhere in the world. This despite the fact that at the time of this writing the hot breath of the California drought could be felt throughout the Okanagan.

The promise of the Okanagan example underscores the fact that we are perhaps one of the few countries in the world that could create a new water ethic. Unlike so many other places, Canada still has slack it can take up in how it manages its water resources. On top of that we actually have a roadmap. By breaking down jurisdictional barriers to integrated water management we could become a sustainable society. And by becoming a sustainable society we have a far better chance of adapting to climate change. If this is what we desire, however, we should be working hard now to break down barriers to enhancement of adaptive capacity while the room to move is still there, for as the final chapter of this book will illustrate, there are many obstacles to the creation of as bright a future as we have just imagined. The most formidable of these include the widespread habits of dismissing or second-guessing science and refusing to act politically in matters related to the environment – except in support of special interests.

As the next chapter will show, a road map will be of no use if we don't choose to follow it. Unfortunately the map this book seeks to create will be no good to us if there is no acceptance at senior levels of government that climate change poses a significant threat to the country's water resources. Without support for the science that is necessary to create baselines for appropriate action, there can be no progress. All we can do is sit and wait to see what will happen.

At the moment, it doesn't appear we are on the road to anywhere nationally. Instead we are driving too fast in the dark without our headlights on, ignoring all the signposts and telling our terrified fellow passengers on this planet that we know where we are going and what we are doing, when clearly we don't. We look to places like the Okanagan and the Northwest Territories to find the road to sustainability again and to direct others to it.

CHAPTER TEN

KNOWLEDGE IS POWER

Building a Better Bridge between
Science and Public Policy

While it existed the Canadian Foundation for Climate and Atmospheric Sciences served as a bridge between the research community and the users of new knowledge in government and in the private sector. It was the main funding body for targeted, university-based research on extreme weather, air quality, climate and marine prediction. It offered a key advantage: its ability to stimulate intersectoral partnerships, particularly between universities and government. The interactions it fostered among government, academic and private sector scientists and managers were all directed toward the development of targeted solutions, but by their nature they also served to enhance Canada's international reputation. When the federal government ceased to fund it, the CFCAS reorganized under the name Canadian Climate Forum.

At the time of this writing the Canadian Climate Forum still exists. It remains a non-governmental agency dedicated to improving understanding of weather and climate in the Earth system. Its mission is to collaborate with partner agencies and individuals to accelerate the uptake and use of weather and climate knowledge to serve the needs of society and the economy.

A SYMPOSIUM ON EXTREME WEATHER

In April 2014 the Canadian Climate Forum held a symposium in Ottawa to examine the incidence and impact of severe weather, the challenges it raises and how to prepare for or adapt to it in order to minimize its often devastating effects. On the evening before the formal symposium opened, an information session open to the public and the media was held at the University of Ottawa. The topics of discussion were the outcomes of the recently released report by Working Group II of the United Nations Intergovernmental Panel on Climate Change Fifth Assessment Report, which outlined the latest research findings related to climate change impacts, vulnerability and adaptation. Discussions also included comments on the IPCC Working Group III report on climate change mitigation. The October 2013 report on the science of climate change was also discussed.

The information session outlined advancements in our understanding of contemporary climate change issues and demonstrated even greater consensus as to the risks associated with failure to adapt to these changes. Over and over again participants in the discussion made the case that climate change is clearly a matter that needs to be addressed far more seriously at the federal level in Canada, and that more harmonization is required between federal, provincial, territorial and municipal governments to accurately assess vulnerability so that meaningful adaptation strategies can be developed that will protect people and their property, critical infrastructure, economic vitality and social and political stability.

The formal proceedings, on April 23 and 24, brought together scientists, agency executives and policy- and decision-makers from the public and private sectors. Presentations and discussions provided insights into extreme events and addressed the economic, infrastructural and health challenges they raise. Participants also discussed measures to develop resilience and to safeguard people and businesses.

In his opening remarks, symposium chair Gordon McBean reiterated that extreme weather events don't just damage property but disrupt lives and alter and stress ecosystems. Extreme events therefore have environmental, health and insurance impacts. McBean noted that we are likely to see 2° to 5°C warming during the current century, and that North America will not be immune. Adaptation and mitigation are therefore necessary. We need leadership, new programs and best practices, McBean said. To highlight these was the goal of the symposium.

After his opening remarks, McBean introduced a former federal commissioner for environment and sustainable development, Scott Vaughan, who gave the symposium's opening keynote. Vaughan began by outlining the contents of the most recent IPCC reports in the context of the 2012 UN report on extreme weather events. The latter report, he said, was important in that it overlaid climate change effects on top of existing vulnerability, something which leaders in the insurance industry had understood for years.

THE KEYNOTE ADDRESS

Mr. Vaughan delivered a powerful and comprehensive assessment of world conditions. There were three main warnings from the recent Fifth Assessment Report. Sea level rise is expected to affect hundreds of millions of people. Higher carbon dioxide concentrations are altering the chemistry of oceans a million times faster than what occurs naturally. Over time this will cut tropical fisheries yields in half, which will have huge impacts on global food security. Vaughan noted that for each 1°C temperature increase there will be a 20 per cent decrease in the amount of freshwater available to a significant part of the global population. This, in addition to growing populations, will demand a 70 per cent increase in food production during this century. Innovative adaptation is beginning but serious challenges remain, particularly in developing countries.

Citing the World Resources Institute, Mr. Vaughan noted that each of the past 36 years has exceeded global average temperatures. The frequency, intensity and duration of extreme weather events are increasing. Heavy rainfalls and droughts are occurring more frequently. Reported claims for flood damages alone have risen from an average of $7-billion a year globally in the 1980s to over $24-billion in 2011. Incremental, linear approaches to adaptation are not enough now and will not be enough in the future.

Risks are clearly rising in Canada, Vaughan said, but we are not moving as quickly as we should to deal with climate change effects. The need to replace much of our aging and inadequate infrastructure must be seen as an opportunity to adapt. Vaughan urged Canadians to look to federal leadership beyond the work presently being done by Environment Canada, Public Safety Canada, Health Canada and Natural Resources Canada in efforts to make the badly needed linkage between risk assessment and risk management. There is a need to mainstream meaningful adaptation approaches in all federal government departments. There is also a need for better understanding of cumulative risks, and more work needs to be done to make these risks known in established engineering practice.

Vaughan noted that there are climate tipping points beyond which sustainable development will no longer be possible for some. We need national leadership and coordination on disaster risk reduction, adaptation and mitigation, he said. We need deep cuts in carbon dioxide emissions. This was not a question of science, he asserted, but of arithmetic, and if we withdraw the subsidies to the fossil fuel industry, the arithmetic will change. Vaughan also commented obliquely on Canada's role in addressing climate risks. If we do not work with the international community to reduce climate effects, we will face a much larger adaptation challenge in the future. "We have got to get moving," Mr. Vaughan concluded.

SESSION ONE: IMPACTS

The Impacts session was opened by the author of the present book, examining the math and aftermath of the catastrophic flooding in Calgary and southern Alberta in June of 2013 and in Toronto less than three weeks later. This presentation drew the attention of participants to the contents of a report prepared by Dr. John Pomeroy for Environment Canada 18 years before, in 1996, which warned of changes in hydrology that would likely increase the potential for flooding, particularly on the Canadian prairies. In that report Dr. Pomeroy had observed that fresh water is both a mediator and a transmitter of climate change effects. Water, he wrote, should be viewed not just as a substance but as a flow of mass, energy and biochemical constituents through and between ecosystems and between the land surface and the atmosphere. Liquid water, water vapour, snow and ice, wrote Pomeroy, transmit climate change impacts across the country and across ecological and political boundaries. This, it appears, is exactly what seems to be happening.

The author's Session One presentation continued with an hour by hour synopsis of hydrologic events during the Alberta flood and compared it to similar disasters that occurred the following July in Toronto, August in Russia and September in Colorado. Focusing back on Alberta, the author pointed out that the flood was not the once in several centuries event many claimed; rather, it was well within what had been experienced in the recent past. This demonstrated that Alberta was not even prepared for the hydro-climatic variability that exists today, he said, let alone what might be projected in the future as the Earth's atmosphere transports more and more water vapour to fuel ever larger floods. The conclusion was that Canada's hydrology appears to be changing. We can expect flood events to be increasingly expensive socially, economically and politically.

The author then offered five lessons that might be learned from the 2013 flood:

1. The loss of hydrologic stationarity could very well be a societal game-changer. What this means is that simply managing water in ways that are useful to us at a local scale will no longer be enough. We have to be alert to changes locally of course, but we now also have to keep an eye on changes in the larger, global hydrological cycle and where possible try to manage and adapt to them. This is a huge new concept, this game changing, and it is going to take time to get our heads around it not just environmentally but economically.

2. What we have seen is what we are going to get. Predicted rises in temperature of between 2° and 6°C would result in further amplification of the hydrological cycle by 15 per cent to 40 per cent or more. This game change is not going to go away. Because of our country's fur trade and colonial history, many Canadian towns and cities are located on flood plains in river valleys. According to a recent survey, some 20 per cent of Canadians believe they live on or near what they describe as a flood plain. Defending or evacuating these areas will be very expensive.

3. The new normal is that there is no new normal. Unless we want our future to continue to be a moving target, sooner or later we will have to confront the fact that we are rapidly altering the composition of our planet's atmosphere, with

significant effects on hydrology – a subject not many political leaders want to talk about in meaningful terms, even after the disastrous flooding of 2013.

4. Structural engineering solutions are going to be necessary, but they are not going to be enough. We cannot ignore the local value of natural ecosystem processes. In order to retain even partial rein over the hydrological cycle we have to enlist all the help nature can provide us. We gain that help by protecting and restoring critical aquatic ecosystem function locally by and reversing land and soil degradation wherever we can.

5. The watershed basin is the minimum scale at which water must managed. This should be perceived as good news. There is much power in realizing that we can do a great deal for our way of life and perhaps sustain our prosperity by taking care of our watersheds. This suggests that it is at the local level – where we live – that we have the most power to effect change and to act most effectively in service of the common good, now and in the future.

In conclusion the author pointed out that this was not the time to throw up our hands in helpless despair. The sky is not falling. The world is not coming to an end. But we do owe it to those who have suffered so much from the 2013 flooding to start getting it right for the next time – for there *will be* a next time – and a time after that. The problems hydro-climatic change is bringing in its wake are not going to go away. Public awareness of water issues is on the rise. There is room to move, he said, and we should get moving while that room still exists.

Dr. John Pomeroy of the University of Saskatchewan was then invited to describe the basin in which the Alberta flooding occurred. High streamflows, he began, are natural in mountain headwaters, and in this context the notion of flooding is a human construct. Pomeroy laid out the synoptic meteorology associated with the storm that brought on the flood. He then showed the weather prediction reanalysis, which demonstrated that it wasn't a wrong weather forecast that affected flood prediction. Rainfall measurements, Pomeroy pointed out, are only part of flood prediction, and in circumstances such as those that created the Alberta flood, only part of the total precipitation is rainfall. Another component in this case was snow, and a great deal of the runoff was caused by warm rain melting the remaining winter snowpack.

Pomeroy observed that snowpacks in the Alberta Rockies were below normal in early May that year. This changed, however, with snowfalls at high altitudes later in the month. Rapid snowmelt contributed significantly to streamflow, which overwhelmed the capacity of already saturated soils to absorb more water.

Dr. Pomeroy went on to describe the research he and his team were undertaking at Marmot Creek in the Kananaskis region of the Alberta foothills, which happened to be epicentre of the storm that started the catastrophic flooding downstream in Calgary. He then graphically illustrated how stream gauges were overwhelmed and then washed away by the torrents created by persistent rainfall and rapid snowmelt on already saturated mountain soils. The potential loss of instrumentation, he explained, is one of the reasons models are so important.

Pomeroy reported that his research had shown that return periods for flood events of the magnitude that occurred in Alberta in June 2013 were about 1 in 32 years at Banff and 1 in 45 years at Calgary. He noted, however, that the floods of record in the past were not caused by the same kinds of storms. Climate change has made the circumstances in which such storms occur different in this place. Pomeroy demonstrated that the region is experiencing warmer winters, with less snow at low elevations and a greater percentage of total precipitation falling as rain. We now have a warmer, drier system throughout the region in which these big storms occur. This, Pomeroy concluded, is why we are witnessing floods and droughts in the same basin in the same year.

Drought is another expression of extreme weather. Ronald Stewart talked about coping with dryness in the context of Manitoba and the Canadian Prairies. After defining drought, Dr. Stewart illustrated its devastating historical impact on the economy of Canada over time. He showed the many factors that influence the occurrence, depth and persistence of drought, and the way wet and dry periods have historically oscillated in Manitoba. This, Stewart explained, was reflected in crop insurance. There were farms in Manitoba, he said, that qualified for crop insurance for both flood relief and drought damage in the same basin in the same year. Stewart then identified drought indicators and various parameters of drought preparedness. He showed how different regions have different susceptibility to drought, and introduced the basic principles behind the proposed surface water management strategy for Manitoba. He also explained how changes in Manitoba Hydro's operations can help mitigate drought impacts in the province.

Stewart then demonstrated drought effects on specific crops. Longer-term droughts, he warned, have to be seen as possible. He went on to describe the challenges associated with predicting where and when droughts will occur, and concluded by noting once again that drought is of great concern in a changing climate, especially as it relates to food security.

Over lunch, renowned Environment Canada meteorologist David Phillips regaled symposium participants with stories of how he chooses his highly celebrated and publicized Canadian Weather Events of the Year. Phillips explained that he bases this wildly popular annual summary of weather disasters in Canada on three criteria: where the event occurred, what happened and how the public responded. Phillips explained that the world has changed over the 20 years he has been offering this annual summary. There are more of us and we're more prosperous, he said, and that is making extreme events more costly. Our landscapes have been re-engineered and there is greater urbanization. Even without climate change, said Phillips, these factors make us more vulnerable to extreme weather events.

Alberta, Phillips said, has been particularly hard hit, qualifying as having experienced the first-, fourth- and fifth-most severe weather events on his long-term list. Alberta is not the only region affected, however. Disasters with insured losses of over $1-billion have occurred every year for the past five years in Canada. Governments and CEOs should be concerned, Phillips pointed out, as extreme events will continue to affect many important social and economic performance measures.

He also noted that extreme weather events appear to cluster together in time and place. We are not experiencing new weather, he said – there are still tornadoes, hailstorms and hurricanes – but the events appear to be statistically larger, more frequent and more intense. In addition to more frequent heavy rainfall, there also appears to be a clear trend toward higher winds. Extreme weather events also cover larger areas and persist longer, making recovery slower and costlier.

Phillips explained that storms that would have been top stories 20 years ago barely rank today, in part because the duration of extreme weather events is lengthening. We are not just breaking weather records, he quipped, we are clobbering them. And it is not the weather we see when we look out the window that is necessarily changing, he said, but the antecedent conditions such as the rainfall or snowfall intensity and soil saturation that may intensify that weather.

Perhaps the biggest weather-related change, Phillips observed, is the concentration of the full effect of extreme weather events on major urban areas. The loss of green spaces in cities, he noted, is clearly contributing to the intensity of such events. The variability of weather over the course of a year is plainly increasing, he said, but so is variability in any given season. "We are experiencing the best times and the worst times all at once," Phillips said. We are witnessing the wettest and driest conditions in the same region in the same season.

In conclusion, Phillips noted that weather disasters are in a growth phase. Changes in these conditions have not slowed; things can clearly get worse; you cannot logically argue it is not happening. If you change the weather, you change the climate, he said, and these trends are not going to go away. "We need to reduce the threat of disaster by preparing for it."

SESSION TWO: EFFECTS ON HUMAN HEALTH

The first afternoon session explored the health impacts of extreme events. Iqbal Kalsi led off, addressing the issue of urban versus rural consequences of extreme weather events as they relate to public health. Kalsi reminded the audience that climate change is the biggest public health threat of the 21st century. Canadians, he noted, should expect greater health impacts as a result of heat waves, with significant associated economic costs. With a projected warming of 2° to 5°c we should expect mortalities to rise.

Kalsi then focused on extreme weather events in the Middlesex-London (Ontario) Health Unit, where he is responsible for the health hazards prevention and management section, which includes climate change effects. In his health unit's district, Kalsi explained, winters are warming and summers are hotter. The frequency of 1 in 100 year, 1 in 200 year and 1 in 500 year floods is increasing, with commensurate public health impacts. There has also been a dramatic increase in tornadoes. The urban heat island effect too is becoming a public health issue in the jurisdiction, and a trend toward higher nighttime temperatures has been observed. Kalsi then demonstrated how these effects will become more pronounced right across Canada.

Groups that will be vulnerable include older adults, infants and young children, farmers and other outdoor workers, the physically active and those with chronic illnesses. Kalsi said his health unit is working with local municipalities to expand the area covered by tree canopy to reduce this effect, but obviously such projects take time.

Constraints to implement programs to reduce the health effects of climate change are many, Kalsi noted, but they all begin with financial resources. The next steps for his health unit, he said, will be to collaborate with stakeholders and current and potential public and private sector partners to create sustainable, healthy communities.

Up next was Dr. Emmanuelle Cadot, to speak about what Canadians might learn from the heat waves that occurred in Europe between 2003 and 2006. Cadot noted that 2003 was the hottest year ever recorded in Europe. It is estimated that in France alone there were 14,802 excess deaths, of which 1254 were in Paris, a 190 per cent increase compared to the reference years of 2000 to 2002. Dr. Cadot went on to demonstrate individual risk factors associated with age, sex, marital status, level of physical activity, state of health and ethnicity. She outlined the Paris program called Chalex (for "chaleur extrême," extreme heat), part of the French government's strategy for information dissemination, prevention and mobilization against elevated-temperature-related climate risks, which includes a national alert system for heat waves that clearly outlines health risks.

Peter Berry of Health Canada's Climate Change and Health office talked about current climate-related impacts on public health in Canada and the preparations that were required to prepare for further changes. Dr. Berry began by drawing attention to a number of heat and other extreme weather events around the world and outlining the human health hazards they posed. These included heat waves, wildfires, undernutrition due to decreases in food production, food and water shortages, vector-borne diseases and mental-health and violence concerns. Berry then spoke about increases in the transmission of Lyme disease in Canada and health impacts associated with the Toronto ice storm of 2013.

Citing what has been done elsewhere, Berry outlined what we can do to prepare for climate change effects on public health in Canada. National public health plans with respect to climate change, he noted, exist in other countries. After demonstrating how preparedness had dramatically reduced mortality from floods in Bangladesh, Berry explained how emergency measures can be aided by clear adaptation strategies in Canada.

Berry outlined a number of specific concerns, including warming impacts on northern Canadians, climate-related spread of infectious diseases, and health impacts associated with extreme heat. Steps forward include heat and health messaging to support increases in personal adaptation; filling knowledge gaps in health-related climate science; identifying factors that increase individual and societal vulnerability; and the development of alert protocols similar to those instituted in France, which will be made available to communities across Canada. In conclusion Berry noted that health-care facility resiliency tools have been developed that include facilitator presentations, resiliency checklists and best practices.

Because of scheduling challenges the final presentation in the health session of the symposium was made on the morning of the second day. Speaking via Skype, Pulitzer Prize winning journalist Sheri Fink described the desperate situation at a hospital in New Orleans in the wake of Hurricane Katrina. In introducing her book

Five Days at Memorial, Fink asked participants of the symposium to give consideration to the impacts of extreme weather events on the most vulnerable in our society and on the people who are entrusted to serve them. Fink invited the audience to think about whether or not, in such extreme circumstances as Katrina, it is even possible to adhere to established moral values or if certain extreme weather made it necessary to abandon such values.

In the aftermath of Katrina, sections of New Orleans were uninhabitable for weeks. The hospital where Fink worked was an island in the middle of the flood zone. When the power went out in the city, the hospital's backup generators could not keep the air conditioning working and still keep the lights on. The vulnerability of those in the hospital increased as hot summer temperatures returned after the storm passed. Helicopters could only take one or two at a time of the 2000 people that needed to be evacuated. The ethical question became the classic dilemma of triage: whom to evacuate first. Do you take the young first because they have a long life ahead of them, or those who because of fragility or age need immediate help without which they will die? When boats finally started to arrive at the hospital's emergency ramp, same hard question: Who first? Another ethical question arose when people arrived at the hospital hoping for medical help or refuge. There were also issues with medical staff who had their own health problems. What do you do when needs outstrip resources? How do you prioritize who lives and who dies? Hurricane Katrina made it clear, Fink explained, that such ethical issues need to be considered in advance.

The situation at Memorial Hospital deteriorated even further when, as feared, the backup generators failed and the lights went out altogether. The question then became whether some patients should be put out of their misery. While that issue was being considered, patients began dying. In the aftermath some doctors and nurses were charged with murder. Ethics and legality, however, rely on intention, and in the case of Memorial Hospital, health workers were not intentionally hastening death but providing comfort. They were found not guilty.

The moral jeopardy that arises in the aftermath of extreme weather events, Fink explained, is similar to what happens in combat zones. As in times of war there are examples of great heroism. In New Orleans during Katrina local volunteers arrived in airboats to evacuate patients. Health workers carried premature babies under their shirts because incubators wouldn't fit aboard helicopters. As in war, Fink concluded, we can't know what we should have done, only know what we would have liked to do.

Better triage, fuller disaster plans and more adaptive preparations are necessary if we are to be able to deal with more frequent, increasingly destructive extreme weather. In concluding her powerful presentation, Sheri Fink pointed to a critical need for political will in recognizing that we are all fragile and vulnerable to the increasing threat of extreme weather events.

SESSION THREE: THE SCIENCE

This segment of the symposium examined specific aspects of science behind our understanding of extreme weather events. Jennifer Francis broadened the scope of the discussion by connecting climate changes in the Arctic with their potential impact on extreme weather events at mid-latitudes. Francis began her presentation with a smorgasbord of wacky weather events from all over the world and asked whether these episodes had anything in common. All of them were related to persistent weather events. The question this raises, said Francis, is whether human-caused climate change is playing a role. She answered by explaining that we have put ourselves in a serious predicament by increasing the concentration of carbon dioxide in the atmosphere. March 2014, she noted, represented the 349th consecutive month with above-average temperatures. She said this meant that if you are presently 29 years old or younger, you live in a climate regime different from the one experienced by people older than you. A warmer atmosphere, Francis explained, is going to provide more moisture to energize and fuel extreme weather events. This will make wet places wetter while increasing evaporation generally, which will make dry places drier.

Arctic sea ice, Francis continued, is now a mere shadow of its former self. There has been a 75 per cent decrease in volume in the past 35 years, and much of what remains is "rotten," or slushy. The year 2012 marked a spectacular loss of Arctic sea ice, which resulted in additional warming of the Arctic Ocean, which in turn melted still more ice. This feedback, Francis explained, causes more and more ice to melt, which is why the Arctic is warming two to three times faster than the rest of the world. This loss of Arctic sea ice is also combining with the diminishing extent and duration of snowcover in the northern hemisphere to result in lessening the temperature difference between the pole and the tropics. Francis believes this has begun to affect the behaviour of the jet stream in the northern hemisphere.

The influence of warmer Arctic air causes the west winds created by the jet stream to become weaker. It is the waviness of the jet stream that creates weather at mid-latitudes. As the jet stream slows and its deviations become wider, weather patterns persist longer and do things we don't expect. When we look at extreme events such as heat waves, heavy snowfalls and long rainfalls we are seeing the result of very large jet stream waves. Are these really getting bigger and/or more frequent? Yes, Francis answered. And this appears to be particularly evident in the North Atlantic in October and December.

She then showed the jet stream patterns for a number of extreme weather events, including extended droughts in parts of the US and the bitter winter experienced in eastern Canada in 2013–2014. Each could be seen to be caused by large waves in the jet stream. Francis concluded by noting that this understanding presents yet another opportunity to demonstrate how human activity is in fact impacting weather patterns in the northern hemisphere

Francis Zwiers of the Pacific Climate Impacts Consortium then weighed in on the question of whether precipitation events have in fact become demonstrably more extreme. He noted there was strong evidence supporting warming temperatures but less clear evidence of changes in precipitation patterns. Zwiers drew attention to the locations of some 8376 weather stations around the world that had been collecting data for 20 years or more. Of these, some 8.5 per cent showed significant measurable increases in rainfall, while 2.2 per cent showed significant decreases. About 64 per cent of these locations did show a correlation between mean temperature and precipitation, however, in accordance with the Clausius–Clapeyron relation, which, as mentioned earlier, describes mathematically how the atmosphere will hold 7 per cent more or less water vapour for every degree Celsius of warming or cooling.

Zwiers went on to demonstrate how computations of potential increases or decreases in precipitation are made in climate modelling. The results of such modelling to date suggest precipitation is intensifying in Canada. What were previously once in 20 year rainfall events now appear to be occurring every 15 years. Zwiers noted that, in his estimation, there is a 25 per cent probability that such events may be attributable to human influence. He went on to cautiously predict that what are now one in 20 year rainfall events could become one in six year events by the end of the century. Zwiers concluded by saying that data limitations continue to hinder clear detection of the climate signal with respect to changes in precipitation. Observed changes, however, do fall within the expectations of the Clausius–Clapeyron relation.

SESSION FOUR: ADAPTATION

In this segment the focus of the symposium turned toward strategies for adapting to extreme weather events in Canada. In the first presentation, Paul Kovacs of the Institute for Catastrophic Loss Reduction talked about adapting Canadian homes to prevent damage from intense rainfall and severe winds. Our houses, he said, are our most valuable asset, and we build good ones in Canada. How, he asked, can we sustain that success?

Kovacs noted that innovative builders are testing and sharing new ideas and slowly enshrining them in building codes and standards. When attempting to bring new approaches to protecting existing houses, however, Kovacs had met considerable resistance. Many Canadian homes, Kovacs noted, are vulnerable to floods and severe winds. Based on insurance claims, it is clear also that severe weather is becoming more common. Insurance claims related to wind and flooding are now greater than claims for damage by fire. As a result, Kovacs said, insurance companies are finding themselves in a new business.

Several factors are increasing risk. These include population growth in areas at risk; aging infrastructure in many parts of the country; and an increase in lavishly finished basements which may be subject to flooding. Precipitation is also increasing. Kovacs's institute put these arguments before builders, and this led to a program that empowers homeowners to consider their particular circumstances through the use of a risk-assessment tool and readily available risk reduction advice, particularly with respect to basement flooding.

The ICLR has also created its own research lab and storm simulator. Research conducted with this simulator has led to adaptation strategies that cost pennies but can save thousands of dollars in damage in the event of high winds or potential flooding. The Institute has also partnered with local governments and the media, through which it celebrates municipal leadership by recognizing progressive jurisdictions and telling their stories.

Kovacs concluded by observing again that we build solid houses in Canada but many of them are vulnerable. The increase in insurance payouts he said, is as much a result of greater vulnerability as it is to more frequent storms. We can build more resilient homes, Kovacs asserted, if we use science as a foundation for action. Progress, however, remains slow.

Deborah Harford of Simon Fraser University's Adaptation to Climate Change Team began her presentation by summarizing the climate change impacts urban areas need to adapt to. She went on to describe the effects of the loss of relative hydrologic stationarity on health, insurance and other important sectors of the Canadian economy. Adaptation, she pointed out, is complicated by urbanization and by the aging of critical infrastructure. Sea level rise is an additional problem for coastal cities such as metropolitan Vancouver. Harford then spoke of the Coastal Cities at Risk Project and its five research themes. These include social vulnerability, health risks, economic exposure, physical hazards and organizational and governance challenges.

Cities, Harford pointed out, are undertaking adaptation projects on their own without provincial support. This has to change, she said. Regional co-operation will ultimately be required to ensure that one jurisdiction isn't undertaking action at the expense of neighbours. The next steps in realizing meaningful adaptation strategies in coastal cities like Vancouver include intergovernmental collaboration; stakeholder engagement, especially with professionals in planning and engineering; continuous assessment of current and future risk; strategic action; and the mainstreaming of adaptive behaviours. Harford concluded with a summary of adaptation needs, goals and barriers. She stressed the importance of updated plans, maps and zoning and of support for municipal adaptation efforts and central leadership.

The next speaker in this session was Don Lemmen of Natural Resources Canada, who talked about adaptation planning in Canada. Lemmen explained why planning mattered and asked rhetorically if extreme weather events shouldn't be used as a wake-up call for advancing adaptation. Noting that risks will affect all economic sectors, Lemmen pointed out the need to more completely link climate adaptation to disaster risk management. These risks, he observed, include immediate effects such as flooding, heat waves, wildfires and droughts but also include slow-onset changes such as sea level rise, glacier retreat, permafrost thaw, ocean acidification and ecosystem changes.

Lemmen pointed out that adaptation is occurring in Canada and that examples of progress abound. Most progress, he noted, has been at the municipal level, which suggests a need for engagement and support from higher levels of government. He went on to describe current and emerging drivers of adaptation, which include concerns about national reputation, access to international markets and matters of regulatory compliance. Lemmen outlined challenges and key regional risks to which extreme events contribute and demonstrated how adaptation can reduce these risks. Increasing exposure of people and assets has been the major cause of increase in disaster risk.

We need to be prepared, he said, and co-operation is necessary, if not critical. Barriers to co-operation have to be identified in advance so they can be overcome when extreme weather events become disasters. We need to ask what practical steps can be taken now, he said, to strengthen the linkages between the climate change adaptation and the disaster risk reduction communities. Lemmen concluded by suggesting that immediately after a natural disaster is a good time to advance adaptation strategies.

Session Four continued with a presentation by Jacinthe Clavet-Gaumont, of the Montreal-based climate research group Ouranos, on the influence of climate change on extremes affecting the hydroelectric sector in Quebec. Clavet-Gaumont began by explaining that 97 per cent of the electricity in Quebec is generated by hydro, which makes reliable distribution highly vulnerable to extreme events. As examples of adaptation, she pointed to the aftermaths of the Saguenay flood of 1996 and the Great Ice Storm of 1998. A process was put into place to identify vulnerabilities and opportunities, develop scenarios and analyze impacts of climate change on targeted activities, with the goal of developing and implementing adaptation strategies.

The development of climate scenarios required downscaling of global models to regional levels and on down to local hydrological models. These models project a 10 per cent increase in the volume of spring flooding in Quebec. They also anticipate that the maximum flood in the most severe of probable meteorological conditions is expected to increase by 28 per cent by 2080.

In summary, Clavet-Gaumont offered three conclusions of note: Hydro-Québec operations are vulnerable to hydro-climatic extremes; hydro power generation and distribution systems are vulnerable to climate change; and therefore adaptation is required.

The final presentation in the adaptation segment was offered by Michel Girard of the Standards Council of Canada, who spoke about the use of standards as vital tools for adaptation. Girard began by introducing participants in the symposium to the Joint Declaration on Conservation of 1908, which more than a century ago concerned itself with many of the environmental issues we face today. In the same way as the Joint Declaration functioned in the early decades of the 20th century, the Standards Council of Canada works today to mainstream adaptation through appropriate standards and verify competence in the achievement of those standards.

Girard pointed out that standards and codes are now being developed through the Northern Infrastructure Standardization Initiative to increase adaptation capacity in order to reduce climate change impacts and disaster risks. For infrastructure these standards include those for design, installation and maintenance of thermo-siphons to keep permafrost from melting; changing standards for snow loads on roofs; management of effects of permafrost degradation on existing buildings; and drainage plans in northern communities that account for the possibility of more-extreme weather events. Girard also described joint US–Canada initiatives on standards for such technologies as balloon-type, ball-shaped backwater valves to reduce basement flooding. He concluded by inventorying various international standardization initiatives and outlining next steps in the exploration of areas of adaptation interest.

SESSION FIVE: IMPACTS AND ADAPTATION IN THE INSURANCE SECTOR

The fifth segment of the symposium dealt with the challenges that extreme events pose to the insurance industry. The first presentation was by Chris White of the Insurance Bureau of Canada, who talked about the financial impacts of extreme weather events on the property and casualty sector.

After noting that the sector was strong and competitive, White outlined the economic contribution of the insurance industry. He cautioned, however, that more-extreme weather events posed a challenge to the entire sector, giving the example that the average basement flood claim in Canada is $50,000. The devastation in Calgary in 2013 was not just dramatic but traumatic, White said. He expressed the concern that another major flood could unbalance the federal budget as it currently stood, possibly impacting the outcome of the next election.

In response to the growing risk associated with extreme weather events, White said, the Insurance Bureau has developed a municipal risk assessment tool which it is piloting in Coquitlam, BC; Fredericton, NB; and Hamilton, Ontario.

The bureau has also done a study on earthquake vulnerability, White added. Canada is the only country in the world that is vulnerable on two coasts, he said, noting that the Insurance Bureau's report had observed that Canadians are not aware or financially prepared for an earthquake of any significant magnitude.

White pointed out that the total economic costs of a major earthquake on the heavily populated west coast were projected to be over

$74-billion, including direct damage to buildings and their contents, business interruption costs and indirect impacts on the economy. This is equivalent to about 4 per cent of Canada's GDP and over 35 per cent of BC's, White observed. Putting this amount in perspective, he said the estimate is also roughly equivalent to the combined output of BC's agriculture, forestry, mining, construction, manufacturing, retail, scientific services and transportation sectors. Insured losses would be much less, he said, totalling $20-billion, and would be largely concentrated in the commercial sector, which could afford and had the foresight to buy earthquake insurance.

For Quebec, White reported, total economic costs of a major earthquake were projected to be over $60-billion, equivalent to 3 per cent of Canada's GDP and almost 17 per cent of Quebec's. Putting this amount too into perspective, White said it would be roughly equivalent to the combined output of Quebec's construction and manufacturing sectors. Insured losses were projected to cover only $12-billion and would be almost entirely in the commercial and industrial sectors, as would be expected given that only 4 per cent of Quebec residents have earthquake insurance.

In conclusion White said individuals have to do more to protect themselves and their property from extreme weather events. Businesses too have to do more, and so do governments. He reiterated his belief that issues of extreme weather events and disaster liability were likely to figure prominently in the next federal election and many elections thereafter.

SESSION SIX: EMERGENCY MANAGEMENT AND PREPAREDNESS

How well prepared are Canada and its communities for extreme weather events? The first presenter on this topic was Ernest MacGillivray, executive director of emergency services for New Brunswick. As he was at that moment dealing with a real-time flooding disaster where he lived and could not leave, Mr. MacGillivray was Skyped into the symposium.

His presentation dealt with where governments currently stood with respect to extreme weather crises. He noted that in a real emergency, disaster service agencies are often seen as independent from government. Emergency services, MacGillivray pointed out, were becoming ever more important in the face of more frequent extreme events. He observed that greater attention needed to be paid to regions where people find themselves repeatedly in harm's way. Risks could be reduced, he said, by better disclosure, by discouragement of development and activities that increase risk, and by appropriate sharing of risks among jurisdictions and governments. This will require a rebalancing of public and private risk, he noted.

MacGillivray described a roadmap for reducing the risk posed by extreme weather events. The roadmap included investments in assessment; fuller disclosure; better public education; and enhanced management of risk that would lead to better control of the outcomes of weather-related disasters. The reason Canadians were not following this roadmap, in MacGillivray's estimation, was largely a consequence of public denial of climate issues. Governments, he said, don't generally lead; they follow. He pointed out that costs associated with disaster relief are already factored into government budgets, and evidence is required to change government direction. Such evidence has to clearly prove that societal and government risks are increasing as a result of more frequent and intense extreme weather events. The federal government, MacGillivray noted, had recently budgeted $200-million over five years, starting in 2015, for a national disaster mitigation program that will be administered by Public Safety Canada. MacGillivray left it to the participants in the symposium to determine for themselves whether this level of funding was adequate.

He then explained how flood disaster mitigation is advancing in New Brunswick. He pointed to a new flood monitoring and alert system and a damage assessment program aimed at helping flood victims almost immediately after an event. But governments can't do what is necessary on their own, he said. What was needed now was an alignment of the work that needed to be done. This alignment includes the implementation of a disaster risk reduction strategy with regular reporting to the provincial government. MacGillivray also noted that there was a need for a new focus on community resilience and sustainability. Greater adaptive capacity also has to be developed at local and regional levels. Building codes have to be altered, land-use plans improved and professional engineering and architectural practice informed by new emergency measures standards.

In conclusion Mr. MacGillivray offered that what we all need to do now is understand the risks, learn from each other, and work to educate government officials and the public so as to better inform policy and practice related to the growing public safety risk of extreme weather events.

The next presentation in Session Six was by Barney Owens, director of response in Ontario's Office of the Fire Marshal and Emergency Management. Owens began by talking about the December 2013 ice storm and showing its effects, noting that at its peak a million people were without power. In such circumstances, he explained, coordination of public communication is central to ensuring public safety.

Owens noted that hydro utilities throughout the affected region immediately shared deployable field resources. As in the case of Hurricane Katrina, however, hospital backup generators proved to be inadequate and had to be augmented with additional units. Small communities proved more resilient during the disaster, Owens said, especially in terms of people helping one another.

He also commented that public expectations range from "it won't affect me" to "the government will immediately come to my aid." One can't count on either of those to be true, and thus personal preparedness is important. As an example, Owens noted that when the power is out, bank machines do not work; in forest fires, cell phone towers burn. He emphasized the importance of working with the media for frequent, accurate communications, as social media can contribute to a loss of control of messaging.

Owens observed that the ice storm affirmed the realization in Ontario that a new emergency measures operations regime is going to be necessary to deal with a changed climate. Ontario learned what constituted vulnerable populations and where they lived. The government also learned that warming centres were critical in such events and that problems with demographics and language can become serious issues in a disaster. The realization also presented itself that there will never be enough money to do the necessary planning. Planning must span multiple jurisdictions and cover both immediate response and the recovery period. Key elements in managing emergencies include promoting personal preparedness; sticking to the plan; applying past recommendations; promoting public/private partnerships; education and awareness; and having sufficient generating capacity where needed.

SESSION SEVEN: THE GLOBAL IMPLICATIONS OF EXTREME WEATHER EVENTS

The final section of the symposium focused on the bigger picture. David Greenall, who leads Deloitte's global climate adaptation and resilience practice, began his presentation by talking about private sector resilience to extreme weather events. He noted there were real risks to business from such events and that these risks may impact enterprise value. Heat waves, for example, may affect cash flow because of lost sales, compromised competitiveness, degradation of critical infrastructure that impacts supply chain reliability, and increased insurance costs.

Greenall then talked about how different responses to extreme weather events can make a difference to both reputation and the bottom line. He gave the example of Hurricane Sandy in October 2012, a storm that knocked out power to over 8.5 million people and caused an estimated $65-billion in damage. Greenall compared the operation of New Jersey Transit with that of the MTA (New York's Metropolitan Transportation Authority) during the storm. In the one case, trains were vulnerable and became unusable, but in the other, a risk assessment had led to operational changes that saw fleet vehicles get moved out of harm's way so that service recovery was fast and financial losses were limited. Similarly, Greenall illustrated what happened in Thailand in 2011, when floods interrupted global supply chains, and in Toronto during the flooding in July 2013.

Greenall explained that Deloitte helps businesses view their profitability through the lens of climate risks, which in this context are modelled using large data sets. Deloitte has encountered some resistance to such assessments, however. Many businesses argue that extreme weather is nothing new or that assessments are unnecessary because such analyses can be done in-house by the company's own engineers. Companies are also wary of committing funds to extreme weather preparedness, as such investments will be lost if threats don't materialize. Greenall explained that businesses need climate and hydrological projections at high resolution. They need to understand direct and indirect impacts and the need for adaptation plans.

Greenall went on to quote a 2013 statement by the American Meteorological Society which predicted that in an increasingly competitive global environment, nations that invest most effectively in clarifying weather and climate risks will have an important competitive advantage. He also quoted Christine Lagarde, the head of the International Monetary Fund, who said in 2013 that "unless we take action on climate change, future generations will be roasted, toasted, fried and grilled."

Greenall concluded his presentation by citing a report by International Finance Corporation called *Enabling Environment for Private Sector Adaptation*, which reviewed actions that have significant potential for enabling private sector adaptation and the promotion of climate resilient development.

SUMMING UP THE COMING STORM

So what, in summary, did we learn from this national conference about where we stand in Canada with respect to extreme weather in the context of climate change? We learned that extreme weather doesn't just damage property. It disrupts lives, impacts economies and alters and stresses ecosystems. Extreme events also have health, insurance and liability impacts. We are likely to see 2° to 5°C of warming over this century, and North America will not be immune. Adaptation and mitigation are therefore necessary.

The recently published Intergovernmental Panel on Climate Change Fifth Assessment Report offers a number of warnings. Sea level rise is expected to affect hundreds of millions of people. Higher carbon dioxide concentrations are altering the chemistry of oceans a million times faster than what occurs naturally, which, in combination with overfishing, will reduce fish stocks. This, in addition to growing populations, will demand a 70 per cent increase in food production during this century. Innovative adaptation is beginning but serious challenges remain, particularly in developing countries.

There will also be significant impacts on water. For each 1°C temperature increase there will be a 20 per cent decrease in the amount of fresh water available to a significant part of the global population. Warming will also cause the hydrological cycle to intensify. The frequency, intensity and duration of extreme weather events are already increasing. Reported insurance claims for flood damage alone have risen from an average of $7-billion a year globally to over $24-billion in 2011. Under these circumstances, incremental, linear approaches to adaptation are not enough now and will not be enough in the future. A storm is coming. We have to get moving.

EXTREME WEATHER EVENTS WILL IMPACT US ALL

Fresh water is both a mediator and a transmitter of climate change effects. Liquid water, water vapour, snow and ice transmit climate change impacts across the country and across ecological and political boundaries. We have to be alert to changes locally of course, but we now also have to keep an eye on changes in the larger, global hydrological cycle and where possible try to manage and adapt to them. This is a huge new concept. "This changes everything," as author Naomi Klein put it.

Floods and droughts represent different sides of the same hydrologic coin, and we have begun to see both occurring in the same basin in the same year. Drought represents a great concern in a changing climate, especially as it relates to food security. Long-term education is needed to encourage drought-proofing and better water management.

There are more of us now and we are more urban, which is making extreme weather more costly. Disasters with insured losses of over $1-billion have occurred every year for the past five years in Canada. In addition to more frequent heavy rainfall, there appears to be a clear trend toward higher winds. Extreme weather events also cover larger areas and are persisting longer, making recovery slower and most costly. These trends are not going away. We need to reduce the threat of disaster by preparing for it.

Climate change has been identified as the single greatest threat

to human health of the 21st century. With projected warming of 2°
to 5°c we should expect mortalities to rise. The urban heat island effect has already become a major public health issue, even in Canada.
Groups that will be vulnerable to these effects include older adults,
infants and young children, farmers and other outdoor workers, the
physically active and those with chronic illnesses. A model worth
emulating has been developed in France for information dissemination, prevention and mobilization against temperature-related climate risks. Included is a national alert system for heat waves that
clearly outlines health risks and shows people how to cope.

Having alert systems in place is one thing; actually dealing with
extreme situations is another. In disasters such as occurred in New
Orleans, it was clearly apparent that it is not always possible to adhere to established moral values. The moral jeopardy that arises
in the aftermath of extreme weather is similar to what happens in
combat zones. There were examples of great heroism. But also as in
war, sometimes it's not possible to know what we should have done,
only what we would have liked to do. In such prolonged emergencies, ethical questions revolve around who gets rescued first. When
needs outstrip resources you may be forced to prioritize who lives
and who dies. Hurricane Katrina made it clear that such ethical issues need to be considered in advance. Better triage, fuller disaster plans and greater adaptive preparation are necessary if we are to
be able to deal with more frequent, increasingly powerful extreme
weather events.

THE SCIENCE IS CLEAR

March 2014 represented the 349th consecutive month with
above-average temperatures. This suggests that if you were 29
years old or younger in 2014, you live in a climate regime different from the one experienced by people older than you. A warmer
atmosphere provides more moisture to energize and fuel extreme
weather events. This will make wet places wetter while increasing
evaporation generally, which will make dry places drier.

Arctic sea ice is disappearing. There has been a 75 per cent decrease in volume in the past 35 years. This loss allows heat to warm
the Arctic Ocean, which in turn melts more ice. This feedback causes

more and more ice to melt, which is why the Arctic is warming two to three times faster than the rest of the world. The loss of Arctic sea ice and the reduction of the extent and duration of snowcover in the northern hemisphere are reducing the temperature gradient between the pole, the mid-latitudes and the tropics. This diminution of the temperature difference between the pole and mid-latitudes has begun to affect the behaviour of the jet stream in the northern hemisphere. It is the jet stream that creates weather at mid-latitudes. As the jet stream slows and its waves become wider, weather patterns persist longer and do things we don't expect.

EXPECT EMERGENCIES

More attention needs to be paid in Canada to areas where people find themselves repeatedly in harm's way. Risks can be reduced by better disclosure; by discouraging development and activities that increase risk; and by appropriate sharing of risks among jurisdictions and governments. Planning must span multiple jurisdictions and cover both immediate response and the recovery period. Key elements in managing emergencies include promoting personal preparedness; sticking to emergency plans; applying past recommendations; promoting public/private partnerships; education and public awareness; and having sufficient electricity generating capacity where needed.

There needs to be a renewed focus on community resilience and sustainability. Greater adaptive capacity also has to be developed at local and regional levels. Building codes have to be altered, land-use plans improved and professional engineering and architectural practice informed by new emergency measures standards.

ADAPTATION

We can build more resilient homes if we use science as a foundation for action. Progress, however, remains slow. Adaptation is complicated by urbanization and the aging of critical infrastructure. Sea level rise is an additional problem for coastal cities.

The next steps in realizing meaningful adaptation strategies include intergovernmental collaboration; stakeholder engagement, especially with professionals in planning and engineering; ongoing assessment of current and future risk; strategic action; and the mainstreaming of adaptive behaviours. Of critical importance are updated plans, more accurate flood maps tied to better zoning, funding support for municipal adaptation efforts and stronger central leadership.

Climate change related risks will affect all economic sectors. There is a need to strengthen the linkages, including institutional connections, between climate change adaptation and disaster risk management. Individuals have to do more to protect themselves and their property from extreme weather events. Businesses have to do more, and so do governments. Issues of extreme weather events and disaster liability are such that they are likely to figure in future elections.

EXTREME WEATHER EVENTS ARE ALREADY IMPACTING BUSINESS

The risks to business globally that are associated with extreme weather events are real and can directly affect reputation, bottom line and enterprise value. The business case for investing in climate resilience is evident. Such investment reduces unwanted exposure and the likelihood of adverse risks with catastrophic consequences; improves business efficiency and effectiveness; proactively addresses growing concerns of regulators, investors, analysts and rating agencies; and creates decision-support frameworks and processes to better equip management to make more informed decisions.

In an increasingly competitive global environment, nations that invest most effectively in clarifying weather and climate risks will have an important competitive advantage.

WHAT DO WE DO NOW?

The problems associated with extreme weather are expected to multiply. The effects of more frequent, intense and longer-duration events such as floods and droughts will have increasing impacts on personal safety, property damage, infrastructure renewal, the reliability of transportation and energy distribution, the cost of insurance, public and private liability, mental and physical health and well-being, economic productivity and social and political stability.

We will all have to adapt. Individuals will need to focus on personal preparedness; communities on resilience and sustainability; and regions on planning and adaptation. Governments at all levels will have to bear down on public education, communication, disaster mitigation, better links to emerging scientific research outcomes, and proactive support for adaptation measures. Professions such as engineering, architecture and urban planning will have to incorporate different parameters into the design and function of our built environment. New levels of emergency preparedness and management will have to be conceived.

Organizations like the Federation of Canadian Municipalities have the opportunity to work with the professional community and with scientists to lead the country's towns and cities in the direction of meaningful adaptation.

Businesses will have to incorporate investments in climate resilience into strategic planning frameworks. The national government will have to invest in both adaptation and mitigation to retain competitive advantage in a world that is about to become very different from the one we knew before our climate changed.

PREPARE FOR THE STORM

In crossing and recrossing this country dealing with water policy matters over the past decade I've discovered a growing sense of unease among informed people that something is brewing out there in the world that is bigger even than the global economic crisis. The sense also is that whatever it is, it is coming at us fast and it has something to do with water. Anybody who is paying attention has already sensed that our basic hydrology is on the move. We don't know what it means yet, but the deep collective sense is that it is not going to be anything good.

A storm is coming and we need to prepare for it. If we do not, then what we will face will be not just any storm but a perfect storm. All perfect storms have one thing in common: when conditions are just right, circumstances can align to create a serial combination of vulnerabilities that lead straight to disaster. We know all about the elements converging in our time to create this tempest. We are fully aware of the global environmental impacts caused by unrestrained growth of human populations and material desire. The converging trade-offs of food security, industrial development, habitat and biodiversity loss and destruction of ocean ecosystems have been expertly described and explained. We have been bickering and prevaricating over what to do about these issues for fifty years, until at last that which many feared most is now upon us.

What is new about climate change is that in synergistic combination with other wide-ranging impacts, warming temperatures threaten to exacerbate already serious problems. Climate change is poised to absorb into itself the hurricane we have already created through population explosion, landscape change and global water scarcity. At the moment, climate change also appears poised to overwhelm institutional structures designed for an earlier, more climatically stable era of human existence. Poverty, deteriorating environmental conditions, jealously protected private property rights, short-sighted economic and market imperatives, self-centred social preferences, unconscionable waste, the increasingly atomized and toxic nature of public discourse, and the reduced capacity of contemporary political frameworks to function in a timely manner in addressing urgent issues are all lining up globally in just the right way to create a hurricane of their own just waiting to be absorbed into the perfect storm of global warming.

If we want to avoid the worst of this storm we need to create a new water ethic. Under the aegis of that ethic we need to ensure formal representation for the environment itself and ways to advocate for nature's own needs for water so as to perpetuate the biodiverse ecosystem productivity that is central to the hydrological conditions favourable to human populations. If we can balance the water availability and water quality needs of nature, agriculture and our cities, many other things we need to do, including achieving sustainability, may very well fall into line. But most importantly, we may be able to avoid turning climate change into a storm many of us might not survive.

ACKNOWLEDGEMENTS

The author wishes to acknowledge the support of the Canadian Foundation for Climate and Atmospheric Sciences in the writing of this book. He would also like to recognize the contribution of Dr. John Pomeroy, whose generous sharing of knowledge and tireless commitment to increasing public awareness of water and water-related climate issues is the inspiration for this book.

This work would not have been possible without the co-operation of all the scientists and others who are named in its chapters, for which I am both honoured and grateful.

The project also benefited greatly from the support of the author's colleagues at the United Nations University Institute for Water, Environment and Health.

· Special thanks are due to Rocky Mountain Books and in particular to publisher Don Gorman for his commitment to books on water and climate; to editor Joe Wilderson for his diligence, intelligence, patience and sense of humour; and to art director Chyla Cardinal, whose mindful designs make books like this shine.

And of course, it must be acknowledged that the author alone takes full responsibility any errors, omissions, misperceptions or misunderstandings the book may contain. Understanding the hydrological cycle and its relation to climate is not easy and is not likely to get any easier any time soon.

A CLIMATE CHANGE BOOKSHELF

Some people change their ways when they see the light,
others when they feel the heat.

— *Carolyn Schroeder*

APPETIZERS: MOVIES

The 11th Hour: Turn Mankind's Darkest Hour into Its Finest. Directed by Leila Conners Petersen and Nadia Conners. Hollywood: Appian Way Productions, Greenhour, Tree Media, Warner Brothers Home Video, 2007. DVD, 95 min.

Global Warming: What's Up With The Weather? Produced and directed by PBS Frontline/Nova. Arlington, Va.: Public Broadcasting Service, 2007. DVD, 112 min.

An Inconvenient Truth: A Global Warning. Directed by Davis Guggenheim. Beverly Hills, Calif.: Lawrence Bender Productions, Participant Media, Paramount Home Entertainment, 2006. DVD, 96 min.

The Refugees of the Blue Planet. Directed by Hélène Choquette and Jean-Philippe Duval. Montreal and Paris: Les Productions Virage, National Film Board of Canada, Point du Jour, 2006. DVD, 54 min.

Six Degrees Could Change The World. Directed by Ron Bowman. Washington, DC: National Geographic Society, 2008. DVD, 96 min.

SUCCINCT SUMMARIES

Emanuel, Kerry. *What We Know About Climate Change.* Cambridge, Mass.: *Boston Review*/MIT Press, 2007.

Saliken, Annette, with Martin G. Clarke. *Cocktail Party Guide to Global Warming: Everything You Need to Know to Converse Intelligently about Global Warming in Any Social Situation.* Victoria, BC: Heritage House, 2009.

STARTING NEAR THE BEGINNING:
BOOKS THAT FRAMED THE DEBATE

Alley, Richard B. *The Two-Mile Time Machine: Ice Cores, Abrupt Climate Change and Our Future.* Princeton, NJ: Princeton University Press, 2000.

Calvin, William H. *A Brain for All Seasons: Human Evolution and Abrupt Climate Change.* Chicago: University of Chicago Press, 2002.

Fagan, Brian. *The Long Summer: How Climate Changed Civilization.* New York: Basic Books, 2004.

Linden, Eugene. *The Winds of Change: Climate, Weather and the Destruction of Civilizations*. New York: Simon and Schuster, 2006.

McKibben, Bill. *The End of Nature*. New York: Random House, 1989.

Ruddiman, William F. *Earth's Climate: Past and Future*. New York: W.H. Freeman & Company, 2001.

———. *Plows, Plagues and Petroleum: How Humans Took Control of Climate*. Princeton, NJ: Princeton University Press, 2005.

Schneider, Stephen H. *Global Warming: Are We Entering the Greenhouse Century?* San Francisco: Sierra Club Books, 1989.

Sherman, Joe. *Gasp! The Swift and Terrible Beauty of Air*. Washington, DC: Shoemaker & Hoard, 2004.

Weart, Spencer R. *The Discovery of Global Warming*. Cambridge, Mass.: Harvard University Press, 2003.

LANDMARKS IN THE OPENING DIALOGUE

Gelbspan, Ross. *Boiling Point: How Politicians, Big Oil and Coal, Journalists and Activists Have Fueled the Climate Crisis – And What We Can Do To Avoid Disaster*. New York: Basic Books, 2004.

McIntosh, Roderick J., Joseph Tainter and Susan Keech McIntosh, eds. *The Way The Wind Blows. Climate, History & Human Action*. New York: Columbia University Press, 2000.

Speth, James Gustave. *Red Sky At Morning: America and the Crisis of the Global Environment. A Citizen's Agenda for Action*. New Haven, Conn.: Yale University Press, 2004.

THE EVIDENCE BECOMES CLEARER

Bowen, Mark. *Thin Ice: Unlocking The Secrets of Climate in the World's Highest Mountains*. New York: Henry Holt, 2005.

Flannery, Tim. *The Weather Makers: How We Are Changing The Climate and What It Means for Life on Earth*. Toronto: HarperCollins Canada, 2005.

Hillman, Mayer, Tina Fawcett and Sudhir Chella Rajan. *The Suicidal Planet: How To Prevent Global Climate Catastrophe*. New York: Thomas Dunne/St. Martin's, 2007.

Kolbert, Elizabeth. *Field Notes from a Catastrophe: Man, Nature and Climate Change*. New York: Bloomsbury Books, 2006.

Lovelock, James. *The Revenge of Gaia: Why the Earth Is Fighting Back and How We Can Still Save Humanity*. London: Penguin Books, 2006.

Monbiot, George. *Heat: How To Stop the Planet From Burning*. Toronto: Doubleday Canada, 2006.

Motavalli, Jim, ed. *Feeling The Heat: Dispatches From The Frontlines of Climate*

Change. From the editors of *E/The Environmental Magazine*. New York: Routledge, 2004.

Pearce, Fred. *The Last Generation: How Nature Will Take Her Revenge for Climate Change*. Toronto: Key Porter Books, 2007.

Romm, Joseph. *Hell and High Water: Global Warming – The Solution and the Politics and What We Should Do*. New York: William Morrow, 2007.

NOTABLE CANADIAN CONTRIBUTIONS

Cruikshank, Julie. *Do Glaciers Listen? Local Knowledge, Colonial Encounters and Social Imagination*. Vancouver: UBC Press, 2005.

Demuth, M.N., D.S. Munro and G.J. Young, eds. *Peyto Glacier: One Century of Science*. Ottawa: Environment Canada, 2006.

Rutter, Nat, Murray Coppold and Dean Rokosh. *Climate Change and Landscape In the Canadian Rocky Mountains*. Field, BC: Burgess Shale Geoscience Foundation, 2006.

Simpson, Jeffrey, Mark Jaccard and Nic Rivers. *Hot Air: Meeting Canada's Climate Change Challenge*. Toronto: McClelland & Stewart, 2007.

Weaver, Andrew. *Keeping Our Cool: Canada In A Warming World*. Toronto: Viking Canada, 2008.

EUROPEAN PERSPECTIVES

Lynas, Mark. *High Tide: The Truth About Our Climate Crisis*. New York: Picador, 2004.

———. *Six Degrees: Our Future on a Hotter Planet*. London: Fourth Estate, 2007.

McDonagh, Sean. *Climate Change: The Challenge of All of Us*. Dublin: Columba, 2006.

THE ECONOMICS OF CLIMATE CHANGE

Barnes, Peter. *Who Owns The Sky? Our Common Assets and the Future of Capitalism*. Washington, DC: Island Press, 2001.

Friedman, Benjamin M. *The Moral Consequences of Economic Growth*. New York: Vintage Books, 2005.

Sachs. Jeffrey D. *Common Wealth: Economics for A Crowded Planet*. The Penguin Press, 2008.

Stern, Nicholas. *The Economics of Climate Change: The Stern Review*. Cambridge and New York: Cambridge University Press, 2006.

CLIMATE CHANGE AND PUBLIC POLICY

Dessler, Andrew E., and Edward A. Parson. *The Science and Politics of Global*

Climate Change: A Guide to the Debate. Cambridge and New York: Cambridge University Press, 2006.

Essex, Christopher, and Ross McKitrick. *Taken By Storm: The Troubled Science, Policy and Politics of Global Warming.* Toronto: Key Porter Books, 2002.

Gore, Al. *The Assault On Reason.* New York: Penguin, 2007.

Kerry, John, and Teresa Heinz Kerry. *This Moment on Earth: Today's New Environmentalists and Their Vision for the Future.* New York: Public Affairs, 2007.

Shearman, David, and Joseph Wayne Smith. *The Climate Change Challenge and the Failure of Democracy.* Westport, Conn.: Praeger, 2007.

ILLUSTRATED BOOKS

Balog, James. *Extreme Ice Now.* Washington, DC: National Geographical Society, 2009.

Braasch, Gary. *Earth Under Fire: How Global Warming Is Changing the World.* With an afterword by Bill McKibben. Berkeley: University of California Press, 2007.

Brown, Paul. *Global Warning: The Last Chance for Change.* Pleasantville, NY: Reader's Digest Assn., 2007.

Collins UK eds. *Fragile Earth: Views of a Changing World.* With a foreword by Sir Ranulph Fiennes. London and New York: HarperCollins, 2006.

Junger, Sebastian, Sir Ranulph Fiennes et al. *Extreme Earth.* With a foreword by Simon Winchester. HarperCollins, 2003.

Knauer, Kelly, ed. *Global Warming: The Causes, The Perils, The Solutions, The Actions: 51 Things You Can Do.* Time Inc. Home Entertainment, 2007.

Ochoa, George, Jennifer Hoffman and Tina Tin. *Climate: The Force that Shapes Our World and the Future of Life on Earth.* Rodale International, 2005.

Schmidt, Gavin, and Joshua Wolfe. *Climate Change: Picturing the Science.* With a foreword by Jeffrey D. Sachs. New York: W.W. Norton, 2009.

Westwell, Ian. *Forces of Nature.* Images from *National Geographic.* Cobham, Surrey, UK: Taj Books, 2008.

CHILDREN'S BOOKS

Bantle, Jason Leo, *Mom, What Can Be Done?* Photographs, with text by Lori Nunn. Christopher Lake, Sask.: Jason Leo Bantle Publishing, 2009.

Okimoto, Jean Davies, with illustrations by Jeremiah Trammell. *Winston of Churchill: One Bear's Battle against Global Warming.* Seattle: Sasquatch Books, 2007.

Strauss, Rochelle. *One Well: The Story of Water on Earth.* Illustrated by Rosemary Woods. Toronto: Kids Can Press, 2007.

DARK VISIONS AND DEPARTURES: NOVELS

Crichton, Michael. *State of Fear*. New York: Harper Collins, 2004.

Kingsolver, Barbara. *Flight Behavior*. New York: HarperCollins, 2012.

Lynch, Jim. *The Highest Tide*. New York: Bloomsbury, 2005.

Tushingham, Mark. *Hotter Than Hell*. Saint John, NB: Dreamcatcher Publishing, 2005.

DARK VISIONS AND DEPARTURES: NON-FICTION

Benton, Michael J. *When Life Nearly Died: The Greatest Mass Extinction of All Time*. New York: Thames & Hudson, 2003.

Diamond, Jared. *Collapse: How Societies Choose To Fail or Succeed*. New York: Viking, 2005.

Homer-Dixon, Thomas. *Environment, Scarcity and Violence*. Princeton, NJ: Princeton University Press, 1999.

Kunstler, James Howard. *The Long Emergency: Surviving the Converging Catastrophes of the Twenty-First Century*. New York: Atlantic Monthly Press, 2005.

———. *World Made By Hand*. New York: Atlantic Monthly Press, 2008.

Martin, James. *The Meaning of the 21st Century: A Vital Blueprint for Ensuring Our Future*. New York: Eden Project Books, 2006.

Tainter, Joseph A. *The Collapse of Complex Societies*. Cambridge, UK: Cambridge University Press, 1988.

Ward, Peter D. *Under a Green Sky: Global Warming, the Mass Extinctions of the Past and What They Can Tell Us About Our Future*. Smithsonian/Collins, 2007.

A BRIGHTER VISION

Homer-Dixon, Thomas. *The Upside of Down: Catastrophe, Creativity and the Renewal of Civilization*. Toronto: Knopf Canada, 2006.

INTERNALLY CONFLICTED NARRATIVES

Lawson, Nigel. *An Appeal to Reason: A Cool Look at Global Warming*. New York: Overlook Duckworth, 2008.

Lomborg, Bjorn. *Cool It: The Skeptical Environmentalist's Guide To Global Warming*. New York: Knopf, 2007.

———. *The Skeptical Environmentalist: Measuring the Real State of the World*. Cambridge and New York: Cambridge University Press, 2001.

BOOKS OF UNCERTAIN VALUE

Joseph, Lawrence E. *Apocalypse 2012: A Scientific Investigation into Civilization's End*. New York: Morgan Road Books, 2007.

Kotter, John, and Holger Rathgeber. *Our Iceberg Is Melting: Changing and Succeeding under Any Conditions*. With artwork by Peter Mueller. New York: St. Martin's Press, 2005.

Solomon, Lawrence. *The Deniers: The World-Renowned Scientists Who Stood Up Against Global Warming Hysteria, Political Persecution and Fraud*. Minneapolis, Minn.: Richard Vigilante Books. 2008.

THE LANGUAGE OF OUTRIGHT DENIAL

Horner, Christopher C. *The Politically Incorrect Guide™ to Global Warming and Environmentalism*. Washington, DC: Regnery, 2007.

Michaels, Patrick J., and Robert C. Balling Jr. *Climate of Extremes: Global Warming Science They Don't Want You to Know*. Washington, DC: Cato Institute, 2009.

Plimer, Ian. *Heaven and Earth: Global Warming the Missing Science*. Lanham, Md.: Taylor Trade Publishing, 2009.

Singer, S. Fred, and Dennis T. Avery. *Unstoppable Global Warming: Every 1,500 Years*. 2nd ed. Lanham, Md.: Rowman & Littlefield, 2008.

Spencer, Roy W. *Climate Confusion: How Global Warming Hysteria Leads to Bad Science, Pandering Politicians and Misguided Policies that Hurt the Poor*. New York: Encounter, 2008.

GOOD BOOKS ON IMPORTANT CLIMATE-RELATED TOPICS

Bakker, Karen. *Eau Canada: The Future of Canada's Water*. Vancouver: UBC Press, 2007.

Economy, Elizabeth C. *The River Runs Black: The Environmental Challenge to China's Future*. Ithaca, NY: Cornell University Press, 2004.

Emanuel, Kerry. *Divine Wind: The History and Science of Hurricanes*. Oxford and New York: Oxford University Press, 2005.

Fink, Sheri. *Five Days at Memorial: Life and Death in a Storm-Ravaged Hospital*. New York: Crown Publishing, 2013.

Fradkin, Phillip. *Wallace Stegner and the American West*. New York: Alfred Knopf, 2008.

Giroux, Henry A. *Stormy Weather: Katrina and the Politics of Disposability*. Boulder, Colo.: Paradigm Publishers, 2006.

Gosnell, Mariana. *Ice: The Nature, the History and the Uses of an Astonishing Substance*. New York: Knopf, 2005.

Hooper, Meredith. *The Ferocious Summer: Adélie Penguins and the Warming of Antarctica*. Vancouver: Greystone Books, 2008.

Isabella, Jude. *Salmon: A Scientific Memoir*. Surrey, BC: Rocky Mountain Books, 2014.

Mooney, Chris. *Storm World: Hurricanes, Politics and the Battle over Global Warming*. Orlando, Fla.: Harcourt, 2007.

Reed, Betsy, ed. *Unnatural Disaster: The Nation on Hurricane Katrina.* New York: Nation Books, 2006.

Safina, Carl. *Song for the Blue Ocean: Encounters Along the World's Coasts and Beneath the Seas.* New York: Henry Holt, 1997.

Saul, John Ralston. *The Comeback: How Aboriginals Are Reclaiming Power and Influence.* Toronto: Penguin Canada, 2015.

Sax, Joseph. "The Public Trust Doctrine in Natural Resource Law: Effective Judicial Intervention," 68 *Mich. L. Rev.* 471–566 (1969–70).

Streever, Bill. *Cold: Adventures in the World's Frozen Places.* New York: Little, Brown, 2009.

Stuart, David E. *Anasazi America.* Albuquerque: University of New Mexico Press, 2000.

Ward, Peter D. *Out of Thin Air: Dinosaurs, Birds, and Earth's Ancient Atmosphere.* Washington, DC: Joseph Henry, 2006.

SURVIVAL; GRASSROOTS POLITICAL ACTION

Clegg, Brian. *The Global Warming Survival Kit: The Must-Have Guide to Overcoming Extreme Weather, Power Cuts, Food Shortages and Other Climate Change Disasters.* Toronto: Viking Canada, 2007.

Dauncey, Guy, with Patrick Mazza. *Stormy Weather: 101 Solutions to Global Climate Change.* With a foreword by Ross Gelbspan. Gabriola Island, BC: New Society Publishers, 2001.

McKibben, Bill, et al. *Fight Global Warming Now: The Handbook for Taking Action in Your Community.* New York: Henry Holt, 2007.

SUSTAINABLE ENERGY

Jaccard, Mark. *Sustainable Fossil Fuels: The Unusual Suspect in the Quest for Clean and Enduring Energy.* Cambridge and New York: Cambridge University Press, 2005.

Scott, David Sanborn. *Smelling Land: The Hydrogen Defense against Climate Catastrophe.* Ottawa: Canadian Hydrogen Association, 2007.

CLIMATE MODELLING

Edwards, Paul N. *A Vast Machine: Computer Models, Climate Data and the Politics of Global Warming.* Cambridge, Mass.: MIT Press, 2010.

McGuffie, Kendal, and Ann Henderson-Sellers. *A Climate Modelling Primer.* 4th ed. Chichester, West Sussex: John Wiley & Sons, 2014.

CLIMATE AND THE ARCTIC

Arctic Climate Impact Assessment. *Arctic Climate Impact Assessment.* Cambridge and New York: Cambridge University Press, 2005.

Byers, Michael. *Who Owns the Arctic? Understanding Sovereignty Disputes in the North.* Vancouver: Douglas & McIntyre. 2009.

Ellis, Richard. *On Thin Ice: The Changing World of the Polar Bear.* New York: Alfred A. Knopf, 2009.

Freeman, Milton M.R., and Lee Foote, eds. *Inuit, Polar Bears and Sustainable Use: Local, National and International Perspectives.* Edmonton: cci Press, 2009.

Koivurova, Timo, E. Carina, E.C.H. Keskitalo and Nigel Bankes, eds. *Climate Governance in the Arctic.* Dordrecht, Netherlands: Springer, 2009

Lord, Nancy. *Early Warming: Crisis and Response in the Climate-Changed North.* Berkeley, Calif.: Counterpoint, 2011.

Struzik, Ed. *The Big Thaw: Travels in the Melting North.* Mississauga, Ont.: John Wiley & Sons Canada, 2009.

CLIMATE AND WATER

Glennon, Robert. *Unquenchable: America's Water Crisis and What to Do About It.* Washington, DC: Island Press 2009.

Murphy, Dallas. *To Follow the Water: Exploring the Ocean to Discover Climate.* New York: Basic Books, 2007.

Schindler, David W., and John R. Vallentyne. *The Algal Bowl: Overfertilization of the World's Freshwaters and Estuaries.* Edmonton: University of Alberta Press, 2008.

Workman, James G. *Heart of Dryness: How the Last Bushmen Can Help Us Endure the Coming Age of Permanent Drought.* New York: Walker & Co, 2009.

2008: THE CLIMATE DEBATE HEATS UP

Atwood, Margaret. *Payback: Debt and the Shadow Side of Wealth.* Toronto: House of Anansi, 2008.

Bowen, Mark. *Censoring Science: Inside the Political Attack on Dr. James Hansen and the Truth of Global Warming.* New York: Dutton, 2008.

Broecker, Wallace S., and Robert Kunzig. *Fixing Climate: What Past Climate Changes Reveal about the Current Threat – and How to Counter It.* New York: Hill and Wang, 2008.

Brown, Lester R. *Plan B 3.0: Mobilizing to Save Civilization.* New York: W.W. Norton, 2008.

Clarke, Tony. *Tar Sands Showdown: Canada and the New Politics of Oil in an Age of Climate Change.* Toronto: James Lorimer & Co., 2008.

Dyer, Gwynne. *Climate Wars.* Toronto: Random House Canada, 2008.

Fagan, Brian. *The Great Warming: Climate Change and the Rise and Fall of Civilizations.* New York: Bloomsbury Press, 2008.

Garvey, James. *The Ethics of Climate Change: Right and Wrong in a Warming World.* New York: Continuum, 2008.

Nikiforuk, Andrew. *Tar Sands: Dirty Oil and the Future of a Continent*. Vancouver: Greystone Books, 2008.

Orlove, Ben, Ellen Weigandt and Brian H. Luckman, eds. *Darkening Peaks: Glacier Retreat, Science and Society*. Berkeley: University of California Press, 2008.

Powell, James Lawrence. *Dead Pool: Lake Powell, Global Warming and the Future of Water in the West*. Berkeley: University of California Press, 2008.

Speth, James Gustave. *The Bridge at the Edge of the World: Capitalism, the Environment and Crossing from Crisis to Sustainability*. New Haven, Conn.: Yale University Press, 2008.

Strom, Robert. *Hot House: Global Climate Change and the Human Condition*. New York: Copernicus/Praxis, 2007.

Zedillo, Ernesto, ed. *Global Warming: Looking Beyond Kyoto*. New Haven, Conn.: Yale Centre for the Study of Globalization/Brookings Institution Press, 2008.

2009: THE LEAD-UP TO COPENHAGEN

Barnosky, Anthony B. *Heatstroke: Nature in an Age of Global Warming*. Washington, DC: Island Press/Shearwater Books, 2009.

Craven, Greg. *What's the Worst that Could Happen? A Rational Response to the Climate Change Debate*. New York: Perigee/Penguin, 2009.

Faris, Stephen. *Forecast: The Consequences of Climate Change, from the Amazon to the Arctic, from Darfur to Napa Valley*. New York: Henry Holt, 2009.

Flannery, Tim. *Now or Never: Why We Need to Act Now to Achieve a Sustainable Future*. Toronto: Harper Collins, 2009.

Gore, Al. *Our Choice: A Plan to Solve the Climate Crisis*. Emmaus, Pa.: Rodale, 2009.

Hansen, James. *Storms of My Grandchildren: The Truth about the Coming Climate Catastrophe and Our Last Chance to Save Humanity*. New York: Bloomsbury USA, 2009.

Hoggan, James, with Richard Littlemore. *Climate Cover-Up: The Crusade to Deny Global Warming*. Vancouver: Greystone, 2009.

Hulme, Mike. *Why We Disagree about Climate Change: Understanding Controversy, Inaction and Opportunity*. Cambridge and New York: Cambridge University Press, 2009.

Lovelock James. *The Vanishing Face of Gaia: A Final Warning*. Toronto: Allen Lane, 2009.

Marks, Susan J. *Aqua Shock: The Water Crisis in America*. New York: Bloomberg, 2009.

Pollack, Henry. *A World Without Ice*. New York: Avery, 2009.

Schneider, Stephen H. *Science as a Contact Sport: Inside the Battle to Save Earth's Climate*. Washington, DC: National Geographic, 2009.

Stanley, John, David R. Loy and Gyurme Dorje, eds. *A Buddhist Response to the Climate Emergency*. Boston: Wisdom Publications, 2009.

Stern, Nicholas. *The Global Deal: Climate Change and the Creation of a New Era of Progress and Prosperity.* New York: Public Affairs, 2009.

Tammemagi, Hans. *Air: Our Planet's Ailing Atmosphere.* Toronto and New York: Oxford University Press, 2009.

2010: AFTER COPENHAGEN

Antholis, William, and Strobe Talbott. *Fast Forward: Ethics and Politics in the Age of Global Warming.* Washington, DC: Brookings Institution, 2010.

Chartres, Colin, and Samyuktha Varma. *Out of Water: From Abundance to Scarcity and How to Solve the World's Water Problems.* Upper Saddle River, NJ: FT Press, 2010.

Cullen, Heidi. *The Weather of the Future: Heat Waves, Extreme Storms and Other Scenes from a Climate-Changed Planet.* New York: HarperCollins, 2010.

Ehrlich, Paul, and Michael Charles Tobias. *Hope on Earth: A Conversation.* With additional comments by John Harte. Chicago: University of Chicago Press, 2014.

Goodell, Jeff. *How to Cool the Planet: Geoengineering and the Audacious Quest to Fix Earth's Climate.* Boston: Houghton Mifflin Harcourt, 2010.

Grant, John. *Denying Science: Conspiracy Theories, Media Distortions and the War Against Reality.* Amherst, NY: Prometheus Books, 2011.

Howe, Joshua. *Behind the Curve: Science and the Politics of Global Warming.* Seattle: University of Washington Press, 2014.

Kennett, James, Kevin Cannariato, Ingrid Hendy and Richard Behl. *Methane Hydrates in Quaternary Climate Change: The Clathrate Gun Hypothesis.* Washington, DC: American Geophysical Union, 2003.

Klein, Naomi. *This Changes Everything: Capitalism vs. The Climate.* Toronto: Knopf Canada, 2014.

Lovelock, James. *A Rough Ride to the Future.* London: Allen Lane, 2014.

McKibben, Bill. *Eaarth: Making A Life on a Tough New Planet.* Toronto: Knopf Canada, 2010.

Ohlson, Kristin. *The Soil Will Save Us: How Scientists, Farmers and Foodies Are Healing the Soil to Save the Planet.* New York: Rodale, 2014.

Oreskes, Naomi, and Erik M. Conway. *Merchants of Doubt: How a Handful of Scientists Obscured the Truth on Issues from Tobacco Smoke to Global Warming.* New York: Bloomsbury, 2010.

———. *The Collapse of Western Civilization: A View from the Future.* New York: Columbia University Press, 2014.

Pearce, Fred. *The Climate Files: The Battle for Truth about Global Warming.* London: Guardian Books, 2010.

Prud'homme, Alex. *The Ripple Effect: The Fate of Freshwater in the Twenty-First Century.* New York: Scribner, 2011.

Rand, Tom. *Waking the Frog: Solutions for Our Climate Change Paralysis.* Toronto: ECW Press, 2014.

Rogers, Peter, and Susan Leal. *Running out of Water: The Looming Crisis and Solutions to Conserve Our Most Precious Resource.* With a foreword by US Rep. Edward J. Markey. New York: Palgrave Macmillan, 2010.

Seidl, Amy. *Finding Higher Ground: Adaptation in the Age of Warming.* Boston: Beacon Press, 2011.

Shearer, Christine. *Kivalina: A Climate Change Story.* Chicago: Haymarket Books, 2011.

Smith, Laurence C. *The World in 2050: Four Forces Shaping Civilization's Northern Future.* New York: Dutton, 2010.

Struzik, Edward. *Future Arctic: Field Notes from a World on the Edge.* Washington, DC: Island Press, 2015.

Sussman, Brian. *Climategate: A Veteran Meteorologist Exposes the Global Warming Scam.* Washington, DC: WND Books, 2010.

Waterman, Jonathan. *Running Dry: A Journey from Source to Sea down the Colorado River.* Washington, DC: National Geographic, 2010.

White, Christopher. *The Melting World: A Journey across America's Vanishing Glaciers.* New York: St. Martin's Press, 2013.

White, Courtney. *The Age of Consequences: A Chronicle of Concern and Hope.* Berkeley, Calif.: Counterpoint, 2015.

White, Rodney. *Climate Change in Canada.* Toronto: Oxford University Press, 2010.

Wilson, Edward O. *The Meaning of Human Existence.* New York: Liveright, 2014.

UN INTERGOVERNMENTAL PANEL ON CLIMATE CHANGE REPORTS

Anyone seriously interested in the climate change issue can readily examine the five major reports published to date by the United Nations Intergovernmental Panel on Climate Change. While the reports themselves are often large and expensive to purchase in book formats, everything the IPCC has published is available at www.ipcc.ch.

GOVERNMENT OF CANADA CLIMATE CHANGE REPORTS

In 2007 the Government of Canada released its own climate change vulnerability assessment, entitled *From Impacts to Adaptation: Canada in a Changing Climate 2007.* This report can be viewed, downloaded or ordered in hardcopy by visiting Natural Resources Canada at www.nrcan.gc.ca/sites/www.nrcan.gc.ca/files/earthsciences/pdf/assess/2007/pdf/full-complet_e.pdf.

The United Nations University Institute for Water, Environment and Health (UNU-INWEH) is a member of the United Nations University family of organizations. It is the UN Think Tank on Water created by the UNU Governing Council in 1996. The mission of the institute is to help resolve pressing water challenges that are of concern to the United Nations, its Member States and their people, through knowledge-based synthesis of existing bodies of scientific discovery; through cutting-edge targeted research that identifies emerging policy issues; through application of on-the-ground scalable solutions based on credible research; and through relevant and targeted public outreach. It is hosted by the Government of Canada and McMaster University.

UNITED NATIONS
UNIVERSITY

UNU-INWEH
Institute for Water,
Environment and Health

Robert William Sandford is the EPCOR Chair for Water and Climate Security at the United Nations University Institute for Water, Environment and Health. He is a water governance adviser and senior policy author for Simon Fraser University's Adaptation to Climate Change Team and is also senior adviser on water issues for the Interaction Council, a global public policy forum composed of more than thirty former national leaders, including Canadian prime minister Jean Chrétien, US president Bill Clinton and prime minister Gro Harlem Brundtland of Norway. Robert is the author or co-author of numerous books on water issues, including *The Columbia River Treaty: A Primer* (RMB, 2014), *Flood Forecast: Climate Risk and Resiliency in Canada* (RMB, 2014), *Saving Lake Winnipeg* (RMB, 2013), *Cold Matters: The State and Fate of Canada's Fresh Water* (RMB, 2012), *Ethical Water: Learning to Value What Matters Most* (RMB, 2011) and *Restoring the Flow: Confronting the World's Water Woes* (RMB, 2009). He lives in Canmore, Alberta, Canada.